edited by Robert Shepard, CAE
graphics by Bert Leveille

AMERICAN FARM BUREAU FEDERATION®

KENDALL/HUNT PUBLISHING COMPANY
4050 Westmark Drive   Dubuque, Iowa 52002

Farm Bureau (FB) is a registered trademark, service mark and collective membership mark owned by the American Farm Bureau Federation.

Copyright © 1996 by the American Farm Bureau Federation. All rights reserved.

Library of Congress Card Catalog Number: 96-78023

ISBN 0-7872-2880-X

All rights reserved. No part of this publication may be reproduced, stored in a retrieval system, or transmitted, in any form or by any means, electronic, mechanical, photocopying, recording, or otherwise, without the prior written permission of the copyright owner.

Printed in the United States of America
10   9   8   7   6   5   4   3   2   1

Introduction

The purpose of this book Your Farm Bureau is to assist county and state leaders and staff achieve a more thorough understanding of the organization. It is meant as an introduction with some tips to help leaders do their jobs more effectively.

The contents of this book was originally written by Alice Sturgis in 1958. The information provided is just as true today as it was when it was first printed. The work of a Farm Bureau leader is a little more hectic and some of the tools available have changed but the basic principals that applied then, still apply today. The book has been edited to bring it up to date in regards to organizational structure and philosophy, but it is basically unchanged.

It is our hope that you the reader will find this a valuable and useful tool as you fulfill your leadership responsibilities within the Farm Bureau organization and for the agricultural community.

> — Robert Shepard, CAE, Director
> Human Resources Development
> American Farm Bureau Federation®

Why This Book

This book is both an analysis and a portrait of Farm Bureau as well as a working guide for its members.

It is important for you, as a member-owner of Farm Bureau, to understand your individual rights, privileges, and responsibilities so that you may enjoy all of them. You also need to be familiar with the procedures which protect your rights and those of your fellow members. It is imperative for you to understand your organization — its philosophy, its strength, its policies, and its plans for carrying them out — so that you may work effectively within its structure. It is vital for you to know how to work cooperatively and efficiently in Farm Bureau. Good members aren't just born: they develop through training, education, and experience.

This book is written to achieve these goals. It represents the philosophy, the faith, and the fundamental convictions of its members. It explains the practical ways by which you may become a more and more valued member of Farm Bureau. It can help you to live and apply Farm Bureau's philosophy of individual freedom and achievement through cooperation with others in your community, your county, your state, and your country.

It offers you the satisfaction of being as effective a member as you are a farmer or rancher.

Table of Contents

1	Chapter 1	You're more important than you think
17	Chapter 2	Why farm bureau is strong
35	Chapter 3	Membership in farm bureau
49	Chapter 4	Majority decision
59	Chapter 5	Procedure in meetings
67	Chapter 6	Business meetings
81	Chapter 7	Officers
93	Chapter 8	Minutes
99	Chapter 9	Discussion in meetings
113	Chapter 10	Voting
119	Chapter 11	Nominations and elections
127	Chapter 12	Committees
137	Chapter 13	Committee reports
143	Chapter 14	County program planning — the process
167	Chapter 15	County program planning — the report
181	Chapter 16	Speakers and audience
197	Chapter 17	How farm bureau develops policies
211	Chapter 18	Carrying out farm bureau policy
227	Chapter 19	Farm bureau and individual freedom
241	Chapter 20	Farm bureau leadership

YOU'RE MORE IMPORTANT THAN YOU THINK

CHAPTER 1

(Photo courtesy Iowa Farm Bureau.)

A young Iowa Farm Bureau family inspects their field of grain as it nears harvest time.

Why You Are Important

You, as a member of Farm Bureau, may not realize how important you really are. Do you fully understand why you are so important both to Farm Bureau and to this country?

When you take a half-hour out of your busy day to plan with a few of your fellow farmers and ranchers on improving a rural road, when you rush through your evening chores and hurry your supper so that you can walk into a Farm Bureau meeting on time, when you get up in a meeting and say what you believe, even though you don't particularly like to talk, you are making a real contribution to self-government. You are more important than you think.

"The greatest phenomena of our American Republic are our voluntary organizations," says one of America's leading industrialists. "They represent the genius of Americans for achieving by working together."

The great voluntary groups, like Farm Bureau, perform functions so important and so unique to this country that they are an inseparable portion of the American way of living. Their greatest contribution is in solving problems – community, state and national problems – not by appeal to government, but by group thinking and united action. Reducing highway fatalities, building a community hospital, playground, or recreation center, securing better schools or roads, promoting trade between countries or peace in the world – all these, and many more, are the constructive jobs which citizens all over America unite in voluntary groups to accomplish.

They do most of their work without recourse to government. They meet the problem themselves. There is scarcely a community which does not profit by the hard work and wisdom of citizens who band together to achieve something of value for their neighborhood, their state, or their county.

You, as an individual member of one of the most fundamental and powerful voluntary organizations, will work with heightened personal satisfaction if you are fully aware of the importance of what you are doing.

You Share in a Trust

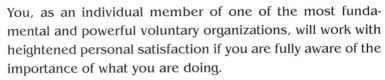

When the membership organization of a profession becomes so well established that the citizens of this country look to it as entitled to speak for the trade or profession, it has, in effect, a trust to represent its particular field in the composite of the nation's viewpoint. For example, the American Bar Association is expected to speak for the field of law and the American Dental Association for the field of dentistry.

Each of the great national organizations, for example the American Farm Bureau Federation, the United States Chamber of Commerce, the National Association of Manufacturers, the AFL-CIO and the American Medical Association, has the high responsibility of studying the problems in its field and helping us decide what is best, not only for the industry, profession, or trade, but also for this nation as a whole.

Each organization must determine the philosophy which guides it, the principles for which it works and the ideas which it seeks to carry out. Each must develop its professional standards; study and seek solutions to the problems peculiar to its profession; educate the public concerning the problems in its field; advise on legislation. Each organization has a trust given it, not only by its own members, but by the citizens of this country to represent and to speak for the nation in its particular field.

Farm Bureau, the general farm organization with by far the largest membership, holds this trust for agriculture. It has, in effect, a vote of confidence, both from its members and from the citizens of this country, that it will execute this trust. To do this demands the best thinking of every member and the highest degree of farmer statesmanship. Each member, each community, each county, each state Farm Bureau has a share of this responsibility. You, as a member of Farm Bureau, are important because you share in this trust for America.

Now Your Voice Is Heard

There was a time when the voice of farmers and ranchers was unheard. They spoke in a confused murmur. Yet they were instinctively seeking association with their fellow farmers and ranchers to discuss their problems and their beliefs.

Sometimes farmers and ranchers would meet at the local cafe to drink coffee, eat donuts and exchange ideas and arguments. "Something should be done," they agreed, as they pondered tariffs, lack of markets, low prices, labor troubles and many other problems. But in the words of one sheep raiser, there was "much talk – and little wool."

Nightly, all over this nation, hundreds of small groups of farm neighbors still gather together. Perhaps there are only six families, perhaps a dozen. They may be around the fireplace or the dining-room table, or out on the screened porch or the patio. Or perhaps twenty or thirty farm families are gathered for their community meeting. But there is a difference. Now they are meeting as an organized group of Farm Bureau. Now, through their own organization, Farm Bureau, they not only talk about the problems – they decide upon solutions and put these solutions into action. Each group is a working unit of a great organization.

Lee, Paul and Jim Eichhorst, Farm Bureau members from Champaign, IL.

(Photo courtesy Illinois Farm Bureau.)

In addition to these small neighborhood and community groups, innumerable county and state Farm Bureau meetings are considering the recommendations which come from these local gatherings. Now your voices are heard, because you speak as one. The opinions of the individual farmers and ranchers who tilted back their chairs in the coffee shop were heard only by a few. The united decision of individual farmers and ranchers today echoes around the world. Through Farm Bureau, farmers and ranchers translate their opinions into decisions, and their decisions into action.

The Magic of Cooperation

The magic which wrought this change is voluntary cooperation through Farm Bureau. Farm Bureau was founded by farmers and ranchers to solve their problems and the problems of their community and their country through voluntary cooperation and self-help. Members of Farm Bureau have proven that the power and possibilities of such a group, trained to work together effectively have no bounds. Teamed up, they can do any job which needs to be done. Farm Bureau preserves the freedom of the individual farmer to think or say or work for what he believes, yet unites the power of all of its members in support of one decision made by a majority vote. Through Farm Bureau, farmers and ranchers make their convictions heard whenever decisions are being made that affect their welfare and the welfare of their fellow citizens.

Individual farmers and ranchers, and neighborhood, community and county groups are, and always will be, most important in Farm Bureau's plan. It is in the individual farmer or rancher and the county Farm Bureau that the strength and power of the organization lie. It is in the local community that ideas are born which, a few months or years later, may materialize in a fertilizer cooperative, a better meat-type hog, a longer staple cotton, or a new community hospital.

You Join Voluntarily

When you and your family join Farm Bureau, you begin to share in this magic of voluntary cooperation. Joining Farm Bureau makes both you and your professional organization stronger. Joining is a voluntary act of decision to work cooperatively with your fellow farmers and ranchers.

You are not drafted into Farm Bureau – you join because you want to: because you believe in association with those who are working for a common cause. You are not denied the right to farm if you do not join. You are not threatened or coerced or compelled. You have no "check off" system. Your dues are not deducted when you sell your grain, your wool, your cotton, or your cattle.

You join of your own free will to have a part in the work of Farm Bureau. After you join you are under no order or compulsion to participate in Farm Bureau activities. You are free to resign at any time.

Your only discipline is the self-discipline of each individual. No one forces you to serve on committees, to vote, or to attend meetings. You gladly contribute thought, time and work because you believe in what Farm Bureau stands for and what it is doing.

The Whole Family Joins

Farm Bureau is a family organization. Unlike the many influences which tend to separate families, Farm Bureau draws them closer together. Fathers, mothers, young people and children work as a family in Farm Bureau. A membership includes the entire family, and Farm Bureau membership is counted by families.

Four year old Johnny plays in the back of the room as his mother attends a policy development committee meeting.

YOU'RE MORE IMPORTANT THAN YOU THINK

Molly and Joan, the sixteen year old twins, delight members at the county annual meeting with their reports of the statewide citizenship seminar which they attended. Big brother Ed rented 80 acres on his own this past summer and is an active member of the county Young Farmer and Rancher Committee. The entire family enjoys the programs, the meetings and the picnic as much as Mother and Dad. Grandma Ellis never misses a Farm Bureau meeting.

YOU JOIN FOR KEEPS

Many of you are fortunate to be born into Farm Bureau. It is one organization in which you plan to remain for life. You join with the intention of being a permanent part of Farm Bureau, and not an in-and-outer. As long as you are eligible you hope to retain your membership.

Fred Walpole lives in a house on the main street in the town of Cross Roads. On his sixty-fifth birthday he turned his farm over to his two capable married sons. He "retired to town" and a chance to travel, which he and his wife enjoy.

As the mailman stopped one morning to hand Fred some letters, he remarked sympathetically, "Too bad you're not a member of Farm Bureau any more, Fred. You'll sure miss those newsletters."

"Who says I'm not a member?" snapped Fred indignantly. "I just mailed my dues yesterday. I may live in town, but I'm a farmer at heart, and farming is paying for my retirement. I'll always be a member of Farm Bureau. I wouldn't any more think of giving up my membership than to stop going to church."

Bill and Ruth Woodman grew up on neighboring farms in Richmond. After attending state college they found jobs in the state capital. Bill works for the state Extension Service and Ruth works as a reporter for the city's daily newspaper. They have maintained their Farm Bureau membership to keep informed on agricultural related issues and they maintain Farm Bureau insurance on their car and their home.

WHO SPEAKS FOR FARMERS AND RANCHERS

Many nonfarm groups claim to speak for them, but their voices are not authentic. Actually, farmers and ranchers can, and do, speak for themselves.

Farmers and ranchers belong to one of the most venerable and most basic professions – agriculture – the producing of food and fiber. Farm Bureau is the voice of that profession – one of the great voices of America. Through Farm Bureau – which you, the members, finance, staff and control – you speak as one.

You speak and vote only after thought and study, and not on impulse. Farmers and ranchers know the importance of allowing time for ideas to mature, for you work with nature, which cannot be hurried.

The voice of farmers and ranchers has distinctive qualities.

The voice of farmers and ranchers is practical. The farmer has his feet on the ground. He has faith in proven natural laws rather than in untried and experimental theories. Always alert to new ideas, he still holds firmly to the fundamentals on which this country was founded and has developed.

The voice of farmers and ranchers is independent. Farmers and ranchers face problems, think and act with the courage of their own convictions and consciences. You are not fearful of upholding an unpopular belief or supporting a little known cause.

The voice of farmers and ranchers is realistic. Farmers and ranchers believe in seeking out and studying all data carefully. You refuse to act on proposals until arguments on both sides have been considered. Your decisions are based not on emotion or wishful thinking, but on facts. You prefer realities to theories.

The voice of farmers and ranchers is constructive. Farmers and ranchers are creative. Fault finding never planted, or raised, or harvested a crop. Farming requires creative ideas as well as hard work. When farmers and ranchers speak, they offer a constructive plan. They are willing to work for that plan and with that plan through its conception, its growth and its fruition.

The voice of farmers and ranchers is heard on things which they believe are fundamentals, and their conclusions are based on long-range thinking. You deal with fundamentals – soil, weather, animals and plants. Your experience teaches you that long-range thinking pays. The farmer who invests in fertilizer, who rotates his crops, who breeds for superior animals five years hence, who plans not for a year but for a generation, practices long-range planning. You apply this same principle to issues, which you study in terms of the fundamental long-range interest of the nation as a whole.

Farmers and ranchers speak with confidence – for they are proud people. They are proud of their profession, proud of their organization and proud of its record. When the American Farm Bureau Federation speaks, it speaks with the united force of several million people. It speaks from knowledge of what its members believe and of what a majority of its members have voted. It has written records of the decisions of its members to back up its statements. It therefore wins the attention of legislators and of the American people.

The advice and guidance of Farm Bureau is sought by our national leaders of both parties. Their confidence in its proposals, its decisions and its policies is evident over the years. The advice asked is not limited to agricultural problems, but extends to a wide range of national and world problems.

Farm Bureau speaks with the authority of an organization which is strong in its thousands of community groups, responsive through its free and democratic procedures, well-informed, not only in the particular interest of its members, but in the needs of the nation as a whole.

page 10   YOU'RE MORE IMPORTANT THAN YOU THINK

Hemstead County, Arkansas, farmer Richard Webb along with wife, Bonnie and 9 month old daughter Lynze, check their cow herd. Webb also has a swine finishing operation where he grows out hogs for Tyson Foods.

(Photo courtesy of Arkansas Farm Bureau.)

So long as each year more and more farmers and ranchers continue to invest their money, their effort, their thinking and their leadership in Farm Bureau, their voices will speak with strength and with unity.

As a member of Farm Bureau you are more important than you think!

JUDGING ORGANIZATIONS

In judging organizations, many people look at superficial characteristics such as:

1. How many "important" people belong to it?
2. How much publicity does it get?

Consultants to voluntary organizations judge them from a completely different viewpoint and rate them by a more fundamental and realistic set of standards. Here are some of the important professional criteria for judging voluntary organizations and the results of these standards as applied to Farm Bureau by the author:

1. **SUITABILITY OF THE ORGANIZATION TO THE PARTICULAR NEEDS OF ITS MEMBERS.** The members of each organization have distinctive needs. Does the structure, and do the activities and services of the organization, meet these needs, or has the group merely borrowed the organization plan of another group?

   Farm Bureau structure is distinctive and it is custom made for the particular needs of its members. It is rare that one finds an almost original organization plan. Farm Bureau is truly a grassroots organization. Farm Bureau originated as county organizations, which then federated into state organizations. This county control gives farm and ranch families unwavering control of their organization.

2. **DEVELOPMENT OF A FUNDAMENTAL GUIDING PHILOSOPHY.** Has the organization worked out a set of basic principles by which all proposals are judged? Or does it improvise whenever a decision is necessary? If it has a philosophy, does that philosophy govern in making decisions?

   Farm Bureau is one of the few organizations that has painstakingly developed and crystallized a practical philosophy. Every proposal of Farm Bureau is judged by its conformity to that philosophy.

3. **CONTINUITY OF PURPOSE.** Do the aims of the organization change with each new set of officers? Does its course zigzag or move steadily forward?

   Farm Bureau's history shows a surprising continuity of purpose and of plan. The organization has chartered a straight course and stays on course.

4. **CAPABILITY OF ITS LEADERS.** Does the organization attract and hold outstanding leaders? Does it offer opportunities for training leaders?

Farm Bureau has unusually capable leaders, some of them internationally known for their contributions to agriculture. It also has leaders in reserve. This did not just happen, however, for the organization consistently works hard at training leaders.

5. **ABILITY OF ITS STAFF MEMBERS.** Does the organization draw and retain capable staff people? Does it challenge their interest and win their loyalty?

Farm Bureau's record attests to the ability of its staff. They remain with Farm Bureau over long periods. Their loyal enthusiasm for the organization is an impressive tribute to Farm Bureau.

6. **DEGREE OF MEMBER PARTICIPATION.** Are decisions made by a few, or do all members participate? Do members do most of the work themselves, or do they leave that to the staff?

I know of no organization which has such complete and ingenious procedures for assuring the participation of all the members. The members actually make all the policy decisions in Farm Bureau.

7. **MEMBER EDUCATION.** How well do the members know their organization? Do they know what it is working for, and why? Are they trained in its procedures and methods?

As a group, the members of Farm Bureau are the best educated of any large professional organization, in all that their organization believes and is working for. They are also thoroughly trained in its procedures.

8. **SERVICE.** How do the achievements of the organization compare with its aims? How important and consistent are its achievements?

arm Bureau rates high in meeting the needs of its members. The program of Farm Bureau arises directly from the distinctive needs of its members. Therefore the members have built a strong program around their needs.

9. **ACHIEVEMENTS.** How do the achievements of the organization compare with its goals? How important and consistent are its achievements?

Every year Farm Bureau lists those things which it hopes to achieve. Some of these are long-range objectives, others are immediate. Consistently each year, the organization checks off a high percentage of its objectives that its members have translated into achievements and selects new objectives.

10. **UNITY OF PURPOSE AND ACTION.** Is there any considerable discord, selfish rivalry, or attempt to use the organization for personal ends?

All of these handicaps to successful cooperation are at a minimum in Farm Bureau. Its leaders learned, in the early years of the organization, that personal gain or advantage must be forgotten if they are to be real leaders in Farm Bureau. Attempts to use the organization for personal advancement have always met with failure.

11. **MAJORITY DECISION.** Do members unite in support of decisions made by the majority?

Farm Bureau's maturity as an organization is proven by the ability of its members, its local groups and its state groups, to unite behind majority decisions. It seeks constantly to improve on an already good record in supporting decisions of the majority.

12. **FINANCIAL STRUCTURE.** Is the organization properly financed? Are its dues or income adequate to provide valuable services? Does it pay salaries that will secure capable staff members? Does it offer members equal opportunity to serve, by paying expenses of officers, dele-

gates and sometimes committee members, while they are doing necessary organization work?

Farm Bureau is soundly and adequately financed. Almost all county and state Farm Bureaus can say yes to each

Morris County, Kansas, Farm Bureau members welcome the public to their corner of the country.

(Photo courtesy Kansas Farm Bureau.)

Mike Bissonette visits with his Jersey cows in Chittenden County, Vermont. His father, Wayne, is President of Chittenden County Farm Bureau.

(Photo courtesy Vermont Farm Bureau.)

of the above questions. Members have been diligent providing adequate financing to accomplish the work of the organization. Measured by any or all professional yardsticks, Farm Bureau is consistently one of the top ranking voluntary organizations. Here is an organization which may be analyzed with great benefit to members of other organizations as well as to its own members.

In 1958, Alice Sturgis stated: "Throughout four years of studying and analyzing Farm Bureau – in national conventions, in board meetings, in staff conferences, in committee meetings, at pot-luck suppers and membership kick-off dinners, in traveling with officers and staff members, and in overnight visits in farm homes over the country – I have been amazed by the ability and performance of Farm Bureau members and staff and by the functioning and achievements of the organization. I have warned myself constantly, they just can't be that good, but further study proves that they **are**!"

That statement is just as true today as it was in 1958. Today's generation of Farm Bureau leaders is just as dedicated and just as skilled. Their achievements are just as notable and perhaps more so, as the percentage of farm and ranch families as a portion of the total population has declined. The challenges are greater and call for a highly skilled group of leaders and staff.

Naturally Farm Bureau has imperfections and its members make mistakes, but the members do not repeat their mistakes and the imperfections are minor.

Farm Bureau is a growing, expanding and changing organization. It faces greater danger of making mistakes than do those organizations which are less active. But Farm Bureau is growing according to a plan, with the constant guidance of an alert and highly skilled membership.

WHY FARM BUREAU IS STRONG

CHAPTER 2

Jo Daviess County Farm Bureau proudly displays its presence on a barn near Elizabeth, Illinois.

(Photo by Robert Shepard, American Farm Bureau.)

Know Your Farm Bureau

Since 1919 Farm Bureau has grown steadily in membership, strength, power and achievement. It is the recognized voice of the farmers and ranchers of America. Farmers and ranchers, educators, legislators and the great bulk of American citizens look to it for guidance in agricultural problems and agricultural planning.

Each organization has an individuality – a personality. Farm Bureau reflects faithfully the personality of its individual farmer members. It is practical, staunchly independent, conservative, constructive, careful in making up its mind, clear thinking, vigorous in carrying out its plans. It looks at problems from a long range point of view, and is thoroughly conscious of the fact that what is good for this country is, in the long run, good for the farmer. Its members are first American citizens and second, farmers and ranchers.

Farm Bureau moves steadily forward, and its membership and its achievements are constantly increasing. To maintain this progress, it is necessary for members to understand many things about Farm Bureau. "Know thyself," meaning "Know Farm Bureau," is the key to working successfully with the group and to using and appreciating its many opportunities and benefits.

Every member needs to understand the answers to these two questions: Why is Farm Bureau strong? From what sources does it draw its strength?

Its strength lies first in its people, its human assets; and second in the working principles of the organization, those ideas which are its intangible, but no less real, assets.

ITS MEMBERS ARE ITS GREATEST ASSET

The first asset of Farm Bureau is its members. You and your son on the next farm – the Ben Browns down the road, the Paul Perkins who own the little farm in New Jersey, the Sam Millers who have a big ranch in Texas, the wheat farmer in Kansas, the dairyman in Wisconsin, the corn grower in Iowa, the fruit producer in California, the cotton planter in Mississippi, the orange grower in Florida, the tobacco planter in Kentucky, the livestock man in Wyoming – these are the greatest and the one indispensable asset of Farm Bureau. Without you and your fellow farm families there would be no Farm Bureau.

An organization is as strong as its members. You, the members of Farm Bureau, united in common cause, have powers and possibilities and potentials which you have only begun to draw upon.

ITS OFFICERS ARE ABLE LEADERS

Farm Bureau's second asset is the officers whom you choose to lead and represent you. You have consistently shown a determination and an ability to seek the best type of leaders. You have demanded and found leaders who have wisdom, courage, patience and practicality, as well as vision, enthusiasm and determination. Their abilities have frequently won national and international recognition.

Farm Bureau leaders make sacrifices – in money, in time taken from their regular work, in absence from family – in order to carry on the work which the members plan. Leadership in Farm Bureau is entrusted to men and women who are dedicated not alone to Farm Bureau, but to the overall welfare of our country. Farm Bureau officers are thoughtfully chosen, trained and tested by experience. They are truly representative of the members and consequently receive their loyal support.

ITS STAFF MEMBERS ARE SKILLED ASSISTANTS

The third asset is the men and women of Farm Bureau's staff. No one understands better than the farmer the importance of securing skilled and devoted assistants to do important work. The staff members of Farm Bureau are fitted by education and experience to carry out Farm Bureau plans and policies. They are as devoted and loyal to Farm Bureau as the most interested members. Their devotion is rooted in a thorough understanding of Farm Bureau and its aims, a genuine desire to serve its members, and a deep founded faith that the members will make the right decisions.

One of America's business leaders made this comment: "In either business or a voluntary organization, the greatest resource is its people – those who work for it. In the case of a voluntary organization, employees play a particularly vital role. As a matter of fact, a voluntary organization cannot succeed unless it has a well organized and dedicated staff.

"Farm Bureau has been spectacularly successful in the achievement of better agriculture in America. That is proof that the member-owners have, and have had, staff members possessing imagination, integrity and devotion."

In addition to its human assets, Farm Bureau has certain basic ideas which serve as its working principles. These human assets, plus these idea assets, are the cornerstones of Farm Bureau's strength. The idea assets are the concepts which Farm Bureau cannot forget or depart from, if it is to grow and to remain strong.

Its Structure Is Sound and Suitable

One of the great working principles of Farm Bureau is maintenance of a structure which is both sound and suitable. The structure of any organization is outlined in the bylaws of its community, county, state and national groups. These are the blueprints from which each group works. They provide the basic plan for the organization's activities. If the structure of a farmhouse is suitable and sound, the farmhouse will be strong and permanent. If the basic structure is poorly planned, unsuitable, or patched together, the farmhouse will not fulfill the expectations of its owners.

The foundation of Farm Bureau's structure is self-government. This foundation upholds a structure which emphasizes the freedom of the individual and the initiation and proposal of policies in the small neighborhood and community groups. The structure of Farm Bureau embodies the whole conception and purpose of the organization – betterment of farm families, community, state and nation by free individuals working voluntarily and together.

This structure enables the members in every community to initiate, consider and decide their own plans instead of following plans handed down by the state or national organization, or by the staff. Every decision is a decision of the members, either by their own vote or by vote of their directly elected representatives. This structure is the incarnation of true self-government.

Farm Bureau structure is unique among organizations. It creates an organization operated by its members – not in name only, but in actual practice. It was created by its members and is maintained and operated by them.

The structure is fundamentally similar to the structure of our government. Like the United States, it is a federation of free and independent member states. Each state has its own particular interests, yet all are united into one great body by their mutuality of interests.

No arbitrary set of bylaws is imposed on the member states, although most bylaws are similar. Each state is bound to the federation by the strongest of all bonds – common interests and voluntary cooperation.

The Resolutions Committee, the most important committee of the American Farm Bureau Federation from the standpoint of policy formulation, corresponds in one important respect to the Senate of the United States. Every state has equal representation on this committee. The smallest state has the same representation as the largest one.

The Voting Delegates of the American Farm Bureau Federation correspond in one important respect to the House of Representatives. Here representation is in proportion to the number of members in each state.

MEMBERS LIVE FARM BUREAU'S PHILOSOPHY

Farm Bureau has developed a strong basic philosophy. It is simple, traditional to America, idealistic, practical. Its source is the Christian faith. Its lodestar is the conviction that freedom of the individual is indispensable to man's well-being and development – a sacred right which comes from the Deity, not as a gift from government.

Farm Bureau members thus reaffirm the faith of our Founding Fathers. Like those pioneers, they fight fiercely in defense of individual freedom. They draw strength from the past and pledge new devotion to an old faith. This dynamic belief in the freedom and dignity of each individual is the motivating force behind Farm Bureau's thinking concerning the rights of the individual, the nature of our government, our economic laws and the functions and responsibilities of Farm Bureau and its members.

Farm Bureau members hold that freedom of the individual is dependent on the upholding of certain basic rights. They believe that every man is entitled to earn money honestly, to save, to invest and to spend his money as he pleases; that his rights to private property must be respected and maintained. They believe that the rights of the individual to speak his mind, to assemble in meetings to choose his leaders, to share in operating his government, to worship as he pleases, are each fundamental to his freedom.

Farm Bureau's philosophy of government follows naturally from this premise of individual freedom. Farm Bureau believes that the highest function of government is to maintain conditions in which individual freedom is stimulated and exercised and in which individual ability and initiative are recognized and can flourish.

Its members believe that the form of government in which each individual earns his reward in proportion to his contribution to society, can best achieve freedom of the individual. In the United States this form of government is commonly called the capitalistic system. They believe that any plan of government which tends to reduce all individuals to the same status, regardless of their capacity or willingness to work – which is commonly called socialism – is destructive of individual liberty. They believe that any step tending toward socialism is dangerous to individual liberty.

They hold that the Constitution is the basic law of this land and that it should be interpreted consistently with the fundamental intent of its founders.

They are convinced that the government must be impartial, that its decisions must be arrived at openly, and that every decision must be based on law. They believe that the powers, duties and regulatory functions of government should be kept to a minimum because centralization and concentration of power can too easily become arbitrary.

They point out that the genius of the Founding Fathers was partly their understanding of the diffusion of power. They divided power among the federal government, the state governments and the people. By similar means they sought to prevent a tyranny of legislative, executive, or judiciary branches of government.

Farm Bureau members hold that it is the duty of government to encourage man's search for a better life by providing constant opportunities for progress; that attempting to direct the decisions of citizens by government propaganda is dangerous to the freedom of the individual and to self-government.

They insist that those who are candidates for office should state openly their beliefs, including those relating to capitalism, socialism and private property rights. All these principles are vital to Farm Bureau's philosophy of government.

The economic philosophy of Farm Bureau also stems logically from its strong belief in freedom of the individual. Its members believe that man can live best under the American system of private competitive enterprise and that supply and demand should be primary determinants of market price. They are convinced that our economic system should be encouraged to produce at its highest rate of efficiency; they advocate the promotion of trade between nations to the mutual advantage of all nations.

Farm Bureau members believe that monopoly in any form is dangerous, whether it be in government, industry, labor, or agriculture; that government regulation ought to be directed toward ensuring the functioning of natural economic laws rather than substituting government decrees for natural laws. They maintain that government regulations should be in the interest of all citizens and that no group should be penalized or rewarded at the expense of any other.

Farm Bureau's philosophy of individual freedom covers not only the individual, his government and his economic life; it extends also to voluntary organizations and to Farm Bureau as an organization. Since voluntary cooperation is a fundamental characteristic of our American system, voluntary organizations are important to the fostering and protection of individual freedoms.

Farm Bureau members believe that government should not participate in the establishment of voluntary organizations or seek to direct them. Farm Bureau members maintain that farmers and ranchers are best qualified to speak for themselves, through the organization of their choice. They accept Farm Bureau as the voice of agriculture. They realize the obligation of the organization to work not only for its own members, but for every citizen; to base decisions not upon what is good for farmers and ranchers, but rather on what is good for this country or for humanity. They apply their philosophy of faith in freedom of the individual to Farm Bureau itself.

Farm Bureau members are convinced that, if America follows the philosophy and vision of its founders, it will remain free and strong in world leadership.

In carrying out its philosophy, Farm Bureau stresses knowledge of facts and common sense. Its members are resolved to know and to use their knowledge to progress. Study, facts and judgement are the tools which they use to work out their philosophy – a philosophy which has proven successful both for America and for Farm Bureau.

Farm Bureau was formed because its members believed that together they could accomplish what they could not do as individuals. Yet the organization continues to emphasize the dignity and freedom of the individual members, to stress the importance of the county Farm Bureau rather than the national organization.

Perhaps this philosophy is the reason why, even in a world torn by strife and clouded by confusion, Farm Bureau members remain calm and march straight ahead toward the achievement of their purposes. Farmers and ranchers have faced the task of developing a philosophy, and they have set it down in plain language. They have faith in it, and they live it.

Its Power Rests in Its Members

A most important working principle of Farm Bureau is that the power of the organization rests in the individual members and in the community groups. It is here that ideas originate. It is here that decisions are carried out. In some voluntary organizations the power of the group rests in a staff which executes plans by utilizing the volunteered time and efforts of the members. In other organizations power lies mainly in a national board or in the national officers. In these types of organizations the members are carrying out the directives of the few in whom power is vested.

The opposite is true of Farm Bureau. Instead of the all too common top down decision and direction, Farm Bureau emphasizes decision and control from the bottom up — control by the members themselves. A few farmers and ranchers in a small community meeting, or working as a committee, can start an idea which will crystallize perhaps a year later in a community improvement. A little prairie community can spark a plan which next year, or five years later, may be a national park or a federal law.

Ideas are born in the little neighborhood groups and in community Farm Bureau meetings. These ideas are studied, discussed and voted upon in these local groups, and then passed on to larger groups of Farm Bureau. After final decision by the members themselves, or by their directly elected representatives, the ideas may become specific projects or policies for Farm Bureau members to carry out.

And who decides state and national policies? You, the members do. When your delegates vote at state and national conventions they vote as your elected representatives. They are there to represent you and they know the decisions you have made at your own local meetings.

You, the members, own, finance, control and operate Farm Bureau. You both create and direct Farm Bureau's power.

Kentucky Senator Wendell Ford addresses Farm Bureau members at their annual meeting in Louisville.

ITS PROCEDURES ENSURE EQUAL OPPORTUNITY

Another principle of Farm Bureau strength lies in its belief in and adherence to procedures which guarantee equal opportunities to each member. Farm Bureau is based upon the same philosophy as our republic; a philosophy of equal individual opportunity to propose measures, to discuss and to decide them. Farm Bureau has developed original procedures and techniques for putting this philosophy into practical use.

Some of the many significant procedures which are distinctive of Farm Bureau include: the initiation of recommendations for policies by individual members and the county Farm Bureaus instead of by the state or national organization; the right of each member to speak and to vote for or against a recommendation or a plan before it is sent to a state or national group; the opportunities for training in member participation and leadership open to every member; the expert aid, advice and education furnished every member by Farm Bureau's officers and staff, its publications, its radio and television programs; and the volume of facts and information about public policy issues available to every member before he or she is asked to make a decision.

Many organizations aim to provide equal opportunities and real control by their members but fail to establish the procedures which ensure that their members can operate and control their organization. Farm Bureau has established procedures which ensure equal opportunity to all.

IT BELIEVES IN ADEQUATE FINANCING

Farmers and ranchers, being practical people, realize that you must spend money to build strength; that an organization which is insufficiently financed soon dies on the vine. Dues are maintained at a level which will permit Farm Bureau to operate effectively and to offer valuable services. In any organization there is a direct relationship between the dues you pay and the services you get. Farm Bureau members believe in a high level of activities supported by adequate financing.

Farmers and ranchers have substantial capital investments in their land, their operating plants, their equipment and their employees. As good businessmen, the owners of Farm Bureau realize the need for maintaining a substantial and continuing investment in their organization, thus ensuring an organization strong enough to promote and protect their business.

IT VALUES EDUCATION

Another fundamental working principle of Farm Bureau is its belief in the value of educating its members in all the ideas and policies for which the organization stands and for which its members are working. The policies and programs of Farm Bureau are based upon thorough study, and upon constant and continuing education, not only of its members, but also of the country as a whole.

Ask any member of Farm Bureau what his organization is working for in any given year and he can usually tell you. He can also tell you why these objectives are important, and how Farm Bureau plans to reach them. The members of many organizations have only a vague idea of what their group is trying to do.

Farm Bureau believes in keeping every member constantly informed about the policies and plans of the organization. It insists on each member having all the facts before making a decision. Farmers believe in lifelong education on all important issues, for better agriculture and better citizenship.

It Has Faith in the Family Unit

Another principle which builds strength is its belief in the importance and value of the entire family working together in Farm Bureau. A Farm Bureau membership includes the whole family. This membership is recognition that each member of the family shares in the business of farming. It is likewise recognition that the fundamental problems of agriculture have no gender.

America's strength lies in her families, and Farm Bureau is an organization in which nine-year old Sally, twelve-year old John, Mother and Dad are all active and interested participants through their family membership. Farm Bureau tends to promote family ties – to keep the family learning, working and having fun together.

Its Members Understand Decision by the Majority

A most fundamental principle of Farm Bureau's strength is a genuine understanding of and adherence to decision by majority vote. Farm Bureau members realize that when they join any organization, they have, in a moral and legal sense, entered into a contract to abide by the decisions of the majority of its members.

Up to the time that a decision by vote is made, every member has the right to oppose a motion or a resolution, to work against it as vigorously as he wishes, and to seek to convert others to his point of view. Once a proposal has been decided by a vote of the majority, however, it becomes the decision of every member of the organization. Regardless of how they voted, they realize that it is now their decision, and they forget their differences and unite in support of that decision.

In the final analysis each organization must speak with one united voice if it is to be heard. No one listens to a babble of conflicting voices. Farm Bureau members understand and work by majority decision. Thus the decisions of the majority of the members become the official policies of all and are supported by all.

IT PROVIDES VALUABLE SERVICES

It is a basic working principle of Farm Bureau that the organization should provide valuable services for its members. Everybody enjoys giving. You are proud to belong to an organization which helps to solve world problems, which initiates programs beneficial not only to local communities and to farmers and ranchers, but to every citizen, and which develops self-help programs in every section of the country.

Members know that the greatest reason for joining Farm Bureau is the opportunity which it offers to unite with other farm families in service to agriculture and in service to our country.

You also enjoy receiving. Farm Bureau provides many services for its members and a wealth of opportunities. The services vary slightly in different states. However, whenever one state develops a new service, other state Farm Bureaus are quick to adopt it also. Thus, most benefits are soon almost uniform throughout the country.

Farm Bureau members have long recognized that it is not what you pay that counts, but rather what you get for what you pay. In general, these are the benefits and services which you, as members, get from Farm Bureau.

When you join Farm Bureau, you...

1. **Obtain** direct, competent presentation of your viewpoint to your county or parish Board of Supervisors, your state Legislature and to Congress, when they are considering legislation in which you are interested.

2. **Ensure** for yourself the opportunity to present your views to the administrative branches of government, through your officers and staff members. Your representatives appear before administrative boards handling such basic matters as transportation rates, equalization of tax assessments and administration of schools. They make your wishes known.

3. **Share** opportunities to explain and clarify farmers' and ranchers' problems to other groups such as sportsmen, businessmen, service clubs, churchmen and union members – thereby promoting mutual understanding, good will, and benefit by informing the public on farm problems.

4. **Gain** the right to propose your own ideas as recommendations in the meetings of your fellow farmers and ranchers; to support your ideas through free discussion; to uphold or oppose the ideas of others; and to vote on all proposals.

5. **Keep** yourself constantly informed and up to date on all matters of interest to farmers and ranchers through the letters, bulletins, magazines, speakers and programs which Farm Bureau furnishes you.

6. **Receive** complete accurate facts and figures for making your own decisions on specific policy recommendations.

7. **Secure** the right to ask the opinions of trained technicians on agricultural problems which concern you. Farm Bureau makes available skilled specialists who prepare up-to-the-minute information for you.

WHY FARM BUREAU IS STRONG

8. **Learn** how to work effectively in any group as leader or member through your experience and training in Farm Bureau. You have constant opportunities for training at institutes, conferences, workshops and conventions.

9. **Save** money, because as a member you can often purchase supplies and services at group discounts.

10. **Qualify** yourself to apply for the best type of insurance at cost. Insurance usually covers life, health, fire, hail, automobile and compensation policies for you and your employees.

11. **Develop** valued friendships for yourself and your family. You enjoy rewarding fellowship with other farm families in all parts of the nation, and opportunities to know the farmers and ranchers in other counties.

12. **Support** agricultural research projects and efforts to inform nonfarm children about agriculture through Ag-in-the-Classroom projects.

13. **Compute** up to the minute price changes in your markets through electronically transmitted market information and learn marketing strategies for the commodities you produce.

WHEN YOU JOIN FARM BUREAU, YOU...

Have working for you hundreds of skilled, enthusiastic and understanding people who are the leaders and staff members of Farm Bureau.

Have working with you thousands of fellow farmers and ranchers of your community, your county, your state and your nation. Farm Bureau is the farmer's and rancher's business, social, educational and government representation organization.

We Believe

- in the American private, competitive enterprise system.
- that the Constitution is the basic law of the land. Long established interpretations should be changed only through constitutional amendment.
- in a government of law, rather than of men and in a Congress that limits discretionary powers of the executive branch and regulatory agencies.
- property rights are among the human rights essential to the preservation of freedom.
- government should provide only minimum aid and control.
- each person should be rewarded according to productive contributions to society.
- government should stimulate, not discourage individual initiative.
- the search for progress should be encouraged through opportunity – rather than hindered by illusions of security.
- monopoly – whether by government, industry, labor or agriculture – is dangerous.
- government should operate impartially in the interests of all.
- propagandizing by government is dangerous to self-government.
- voluntary cooperation is a part of the American system – and is the "Farm Bureau way."

(Photo courtesy Indiana Farm Bureau.)

Owen and Della Menchhofer, left, of Ripley County, Indiana and Janet and Alan Kemper of Tippecanoe County, Indiana give each other pats on the back during an American Farm Bureau convention.

(Photo courtesy Kentucky Farm Bureau.)

Henry County, Kentucky, Farm Bureau membership recruiters never got over the fence but they made the sale. From left, the Farm Bureau contingent is Mark Lyle, Leo Lyle and county president John E. Smith. They are talking to recruit Kevin Flood.

Membership in Farm Bureau

CHAPTER 3

Waseca County, Minnesota, Farm Bureau membership chair Steve Scheffert, at left, signing up "new member" Scott Hildebrandt.

(Photo by Joan Waldoch, American Farm Bureau.)

Rights of Members

An individual who takes pride in his profession, occupation, or skill naturally seeks membership in the organization of his coworkers. When you join Farm Bureau, or any other voluntary organization, the law says that, in effect, you establish a contract between yourself and the organization. Under that contract you acquire certain fundamental rights; these rights may be changed or broadened from time to time by amending the provisions of the bylaws.

When you join Farm Bureau you acquire the legal right to share in all of the benefits, opportunities, privileges and advantages enjoyed by its members.

Among the specific membership rights which voting members acquire are the following:

1. To be notified of meetings
2. To receive official Farm Bureau publications
3. To attend meetings
4. To present motions or resolutions for consideration by Farm Bureau members
5. To discuss questions at Farm Bureau meetings and to advocate and work for whatever action you think best
6. To vote
7. To nominate candidates
8. To be a candidate for Farm Bureau office
9. To consult official records of the organization
10. To insist on the enforcement of the rules and procedures of Farm Bureau
11. To share equally in all benefits offered by Farm Bureau.

OBLIGATIONS OF MEMBERS

In return for these rights of membership, you assume certain obligations and responsibilities. Your most important obligations are:

1. To study and vote on Farm Bureau issues
2. To participate in developing Farm Bureau policies
3. To abide by the decisions of the majority of the members
4. To support Farm Bureau policies
5. To carry out duties which may be assigned to you
6. To work within the structure of the organization and according to its policies and rules.

You as a member must face and undertake your share of the responsibilities which you assume when you join Farm Bureau. You can't shift or dodge or bypass the responsibilities of membership. They are obligations which must be met if you, as a member, and Farm Bureau, as an organization, are to advance.

If your county Farm Bureau votes to secure the improvement of Centerville Highway, you have both a moral and legal responsibility to do your share of the work involved. Do you help to get signatures, do you assist in gathering the facts which will prove the need for the better highway, do you attend the hearings on the proposal?

Or do you try to shift your share of the responsibility to "somebody who isn't so busy" or "to the committee," or to "John Falk, who proposed the idea, anyway?"

Working on a project and in an organization is like rowing on a crew. If one oarsman fails to do his share of the rowing, the whole crew is slowed down and the balance and power of joint effort is upset. The man who stops rowing and rests on his oars can't shift his responsibility to the rest of the crew. Nor can they take his responsibility. When you join a team or an organization, you have assumed a responsibility which only you can fulfill.

Working within the structure of an organization means working through the channels which have been established by the vote of you and your fellow members. For example, if you wish to present an idea which concerns Farm Bureau, you do it in the manner agreed upon, by presenting it first to your community group – not by telegraphing the governor or by giving an interview to the newspaper. Thus you observe the procedural pattern of your organization and work within its structure.

This contractual relationship of rights and obligations continues until ended by you or by the Farm Bureau.

Women of Farm Bureau

Membership in Farm Bureau is counted by families and every member of the family shares in these joint rights and obligations. The women of Farm Bureau are regular, full-fledged members and participate in all of its programs and activities.

There are projects and programs with which women particularly enjoy working, and the Women's Committees of Farm Bureau shoulder responsibility for these projects as well as doing their share of regular Farm Bureau work. They also assume full responsibility for certain specified programs, at the request of Farm Bureau.

Farm Bureau women, unlike the women in many other groups, are not auxiliary members – they are regular members. They are integrated and represented throughout Farm Bureau's entire structure.

The Women's Committee of the American Farm Bureau Federation is made up of a chairman, a vice-chair and two members from each of the four regions into which the nation is divided. All of the members of the committee are elected by representatives of the state Farm Bureau Women's

Committees. The Chair of the Women's Committee is a member of the Board of Directors of the American Farm Bureau Federation.

The Women's Committees of most of the states are similarly organized and usually the chair is a member of the state board of directors. The women contribute generously to all activities of Farm Bureau. They carry their share of membership responsibilities just as they do in the farm business. Their potential strength and abilities are high. Their strength and their abilities can be increasingly developed and utilized with tremendous benefit to Farm Bureau.

In Farm Bureau, women are working in productive relationships with men, young adults and children – just as they work within their families and in the operation of the farm. As wives and mothers in the family group, they have learned tact and understanding in working with other members of the family. They are able to project this experience and tact in working within the larger Farm Bureau family.

In the Farm Bureau family, men, women, young adults, all have learned to work for the good of the whole group. They know that ideas and policies are decided on their own merits, not by who sponsors them. They know that officers are chosen, committees are appointed and responsibility is delegated, not on personal basis, but on individual ability.

Many women serve in leadership positions within the Farm Bureau. Most counties have women serving on their county boards and committees and in many cases a woman serves as county president. Women serve on state Boards of Directors, elected in their own right. Women currently serve as state Farm Bureau presidents. Women represent the organization before legislators and congressmen and in meetings with administration officials. Their influence and contributions grow with each passing year.

Farm Bureau women are full partners in the business of farming and full partners in the Farm Bureau family.

Young Farmers and Ranchers of Farm Bureau (YF&R)

The Young Farmers and Ranchers of Farm Bureau are also an integral and valuable part of all the organization's activities. In addition, these young adults have developed certain activities and programs of their own choosing, suited to their own needs, and conducted by their own members, through the Young Farmers and Ranchers Committees of Farm Bureau.

Farm Bureau supports and encourages these activities . The organization realized that young people who are given opportunities for training assume responsibilities early. Farm Bureau needs well-trained young leaders. It needs their idealism, enthusiasm and courage. The young people need Farm Bureau – its support, its experience and its counsel.

The YF&R groups provide experience in working on committees and in planning and participating in meetings. They are the laboratories where young Farm Bureau members learn the rules and perfect the art of working together and gain experience in working as a team. Then as trained leaders they move on into even more responsible positions with Farm Bureau.

Farming as a profession and Farm Bureau as an organization are becoming more complex. Young leaders and young members therefore need more training, so that they can hold the advances and achievements of older leaders and members and continue to move forward to still greater accomplishments.

The young leaders of Farm Bureau can and do contribute mightily to Farm Bureau's strength, and they assume a constantly heavier load. All are workers and many are leaders in today's job of producing a better agriculture.

Farm Bureau members know that, in the long-range planning for future progress, the young people are the most important members of the Farm Bureau family.

Participation Pays Dividends

Joining Farm Bureau means a personal decision to affiliate yourself with the fellow members of your profession. Your benefits are proportional in part to the investment of time and effort which you make.

To realize the full benefits of Farm Bureau membership one has to participate actively in the group. A member who is proud to accept assignments to committees, quick to volunteer when workers are needed, and regular in attendance at meetings gets the most from his organization. Farm Bureau membership is doubly rewarding to the active participating member.

Members cannot all make the same or equivalent contributions to Farm Bureau. The minimum contribution is becoming a member. This is certainly preferable to ignoring the official membership organization of your profession, yet being willing to profit by its efforts. To support your professional organization by becoming and remaining a member of it, shows good will and gives a real backing.

Participation cannot always be on a continuous, active basis; sometimes a member gives generously of his time for years and then, because of other obligations, remains comparatively inactive for a time.

Members of the Maine Farm Bureau Women's Committee, Elaine Beal and Karen Price, give one another their support and a big hug.

(Photo courtesy Maine Farm Bureau.)

Tests of a Good Member

You are a valuable member of Farm Bureau if you give the best that you can to your organization of your time, your interest and – most important of all – your loyalty.

These questions may help you to determine for yourself how good a member you are, as judged by the usual standards of membership:

1. **DO YOU UNDERSTAND COOPERATION?** Most of us are born individualists. We have to learn to exercise self-discipline before we can work cooperatively with a group.

   Real cooperation requires that each member think first of the interests of the group as a whole and only secondarily of his own personal interests. Cooperative achievement is possible only when there is open-mindedness and a determination to work for the good of all rather than for the good of a few.

   To work cooperatively you must be a good sport. You may run for office and be defeated, your pet suggestion may be ignored, your motion may be voted down. But your loyalty does not falter and you don't sulk. You remind yourself that competent groups make decisions, not on a personal basis, but on the merit of an idea.

   Mr. A may be the best-liked member in the whole county Farm Bureau. Yet a motion proposed by Mr. A may be soundly defeated. This is because thinking people vote on the value of an idea, not on their feeling toward its sponsor.

   The power of any group working together is far greater than the total of the power of the same individuals working separately. Voluntary cooperation can work miracles. Its possibilities are limitless.

2. **DO YOU ENCOURAGE OTHER MEMBERS?** Or do you take the work that others do for granted? Do you

remember to congratulate the chairman who is responsible for an excellent program, the speaker who has offered a constructive thought, and the committee chairman who has presented an exceptional report?

Compliment the secretary when his minutes are clear and concise, and the speaker if you enjoyed his talk. Don't forget to show appreciation to your fellow members, too. For example, tell him, "Bill, you deserve a medal for driving thirty miles on a night like this to get to this meeting." Showing appreciation costs you nothing, but it makes you and the other fellow feel good all over. Everyone loves a member who encourages others.

3. **DO YOU ASSUME RESPONSIBILITY WITHOUT BEING ASKED?** For example, if someone who is not a member misunderstands what Farm Bureau is doing, do you courteously set him right on facts?

If there aren't enough chairs at a meeting, do you bring in more chairs? If you see visitors sitting alone, do you chat with them even though you are not on the hospitality committee? You don't need to be a "take-charge guy" to perform valuable services when they are needed.

Do you try to get a new member whenever you meet someone who is eligible, or do you "leave that to the membership committee?" Do you wait for someone to ask you to take responsibilities, or do you volunteer wherever you see that you can be useful?

4. **DO YOU OFFER CONSTRUCTIVE SUGGESTIONS?** The member who can offer helpful ideas builds up and stimulates the group. It is easy to criticize adversely and to find fault. It takes neither skill nor brains to sit on the sidelines and grumble. On the other hand, it requires thought and intelligence to offer sound helpful ideas and plans. If you do find fault, it is your duty to suggest a remedy.

If you are disappointed in your community Farm Bureau meetings or in the program for the year, don't turn your back and walk off, mumbling "I'm staying home until they get better programs." Set to work to improve the programs, to supply the needs which are unfilled, to offer constructive ideas.

5. **DO YOU USUALLY SERVE ON COMMITTEES IF APPOINTED?** Committees do most of the work of Farm Bureau. It is through committee work that you become well-acquainted with other members and grow to understand your organization intimately. If you are appointed it is because your fellow members have confidence in what you can contribute. If you actually can't take a committee job this year, perhaps you can plan to accept one next year.

6. **DO YOU ATTEND MEETINGS REGULARLY?** The power and the productivity of Farm Bureau is created and nourished in its meetings. The members who attend are the ones who propose, discuss and decide important issues.

You may sometimes feel that your absence doesn't make much difference. "I'm just one member. They'll never miss me." But suppose every other member had the same feeling. There would be no meeting. Every member who attends is a contributor.

A member who fails to attend a meeting, or who leaves before adjournment, has no right to question or criticize the actions of that meeting. By his absence he has delegated his share of decision-making to those who are present.

7. **DO YOU SAVE FARM BUREAU TIME AND MONEY WHENEVER YOU CAN?** For example, do you wait until you get a bill for your Farm Bureau dues, or do you send a check promptly at the beginning of the fiscal year?

No farmer forgets to pay his taxes – if he does, he pays a penalty. Similarly, the Farm Bureau member who forgets to pay his dues promptly also pays a penalty – he

MEMBERSHIP IN FARM BUREAU

wastes the time of staff members and the time and energy of his neighbors who call on him to collect his dues.

If you are temporarily short of funds, do not hesitate to send a note to your community chairman or your county treasurer, explaining why your dues will be delayed. You will meet with understanding.

Every member hopes to remain in Farm Bureau for life. Membership is not an off-again-on-again matter. A man who joins the church of his choice, the Masons, the Knights of Columbus, or the Elks expects to be a member for life, just as you expect to remain a member of Farm Bureau as long as farming is your chief interest and as long as you can qualify. So save Farm Bureau time and money every way you can.

Do you pass the test for a good member as judged by these usual standards?

Brad Phillips, a county director from Maine, speaking at the Maine state fair.

(Photo courtesy Maine Farm Bureau.)

Each in His Own Way

Frequently Farm Bureau members give gloriously in unique ways. Even members who are handicapped by illness or other circumstances make some of the greatest gifts. Here are a few examples:

3 MEMBERSHIP IN FARM BUREAU

In the Northwest one poultry farmer, with a large family and a small income and no time to serve on committees, happens also to be an excellent musician. Busy as he is, he leads the singing whenever he is needed – in his community center, at the county annual meeting and at conventions. He radiates enthusiasm, and Farm Bureau members love to sing with him. Many a meeting has "gone over" because he gives generously of his talent.

In a small community on the Atlantic coast there is a farmer's wife who must stay at their dairy at night, so she seldom gets to meetings. She is an appreciated contributor to Farm Bureau. She reviews books for the county paper, and each book she reviews is hers to keep. She gives these to her community Farm Bureau and thus has built a circulating library which she enlarges at the rate of six to eight books a month. Another member acts as librarian. The continuous circulation of these books is proof of the value of what these two women are doing.

In one Midwestern state there is a member who would "rather be shot" than stand up and talk publicly. He is just plain bashful and never yet has said a word in a Farm Bureau meeting. But he is counted one of the most helpful members of his community Farm Bureau. He recently presented the local 4-H Club with a purebred heifer. Each year the Young Farmers and Ranchers committee uses one of his timbered pastures bordering a large lake for a week-end camp. He goes to considerable expense in furnishing facilities and equipment for cooking.

There is a farmer's wife near Lake Michigan who is completely deaf, and hearing aids do not help her. As a consequence she doesn't enjoy meetings, but she loves Farm Bureau. She makes the best banana cream pie in the state – several prizes at the state fair prove this. There hasn't been a potluck supper, or a Farm Bureau Women's luncheon or even a county banquet in her part of the country for a long time where she did not supply pies. Her pie donations run into the hundreds, and the membership chairman was heard to surmise that he thinks "a lot of people join Farm Bureau partly to eat those banana cream pies."

In New England there is a member who runs his farm from a wheel chair. He is crippled and never leaves the home place. "I can't come to you – so you come to me," begins his invitation for the annual county Farm Bureau picnic which is always held on his farm. This picnic is eagerly anticipated from year to year, for his Class A dairy supplies not only the cream for the coffee and the butter for the biscuits, but gallons of his famous home-made ice cream.

On the West Coast a young man and his widowed mother run a dairy. The son has little time for meetings. His contribution to Farm Bureau is unique. Each year he organizes and sets up a booth at the county fair where his YF&R Committee rents wheel chairs for the aged and strollers to parents who have brought children. Last year the booth netted $380 for Farm Bureau.

Near the foothills of the Ozark mountains lives a farm family of three. The wife cares for her bedridden mother who cannot be left untended. On meeting nights the husband usually goes alone. Occasionally, he stays at home with the invalid and his wife goes to the meeting instead. Yet each individual of that family has made substantial contributions to their community Farm Bureau.

When the members decided to restore the old schoolhouse for a Farm Bureau meeting place, the family talked over what they might do to help. "I know," said the wife suddenly, "I'll design a new pattern for a bedspread and we'll weave some spreads to sell for Farm Bureau. Mother can help wind the yarn and..."

"And I'll take them into Bill Cummins and ask to show them in his store window," interrupted her husband. "We'll all three have a share in helping." This family's project cleared over $400 toward the new meeting place.

MEMBERSHIP IN FARM BUREAU

The hobby of one truck farmer is woodcarving. He has a beautiful stand of maple trees on his farm. Last year, though shorthanded because his two sons were in the Army, he turned out a handsome gavel for the Farm Bureau chairman. He worked on it in spare moments and in the evenings. This year he presented Farm Bureau with a speaker's stand, complete with a light. This lectern is greatly appreciated, not only by the members, but by visiting speakers. Now he is at work on a tray which he plans to give to whoever brings in the most new members during next year.

Not every member can attend meetings. Not every member can serve on committees. Not every member can get up and give speeches or devote a lot of time to Farm Bureau. But **every member can help in some way.** We can all contribute – each in his own way.

To the member who works out a way of contributing to his organization by using his particular abilities or assets in his own way, comes a lasting and profound satisfaction. He is proud of his plan to help and he pursues it vigorously through the years.

MAJORITY DECISION

CHAPTER 4

Michigan Farm Bureau's Seventy-Fifth Anniversary Annual Meeting was held at the Westin Hotel and Renaissance Center in Detroit. Michigan Farm Bureau President, Jack Laurie presides over the delegate sessions.

(Photo courtesy Michigan Farm Bureau.)

What Decision by the Majority Means

"The vote of the majority decides."

We have heard this statement since childhood because we live in a country which practices self-government. Our government is founded upon decision by majority vote. But sometimes we do not understand the real meaning of familiar phrases. Or perhaps we have heard them so often that we have forgotten their meaning. What do we mean when we say: "The vote of the majority decision?"

Why not a two-thirds, or three-fourths, or unanimous vote instead? Is decision by a majority vote vital in principle, or is it just a convenient number or an established custom?

Decision by majority vote is the foundation of self-government. Thomas Jefferson recognized this truth. In 1817 in a letter to a friend, he explained his conviction that:

"The first principle of republicanism is that rule by majority decision is the fundamental law of every society of equal rights; to consider the will of the society announced by the majority of a single vote, as sacred as if unanimous, is the first of all lessons in importance, yet the last which is thoroughly learnt."

Jefferson is saying that decision by majority vote is the most important principle of every organization in which members have equal rights; that decision by a majority of one vote – for example a vote of 17 to 16 – is binding and as sacred as a unanimous vote. He warns that though this lesson is the most important of all lessons, it is usually the hardest to learn and the last to be understood. One evidence of a member's experience and maturity is his understanding of majority decision, and his ability to practice it in fact and in spirit.

When you join the Farm Bureau or any other voluntary organization, in the eyes of the law you enter into a contract to do certain things. One of the things you agree to do is to abide by the decision of the majority of the members of the organization.

MAJORITY DECISION

When any question comes before your committee meeting, community meeting, county meeting, state or national convention, you have a right to oppose it by speaking from the floor as vigorously as you wish. You have the right to campaign and to try to persuade others to vote with you on the question.

If it happens that the majority decision is against you, your rights of opposition cease, at least for the present. The moment the vote is taken, the majority decision becomes your decision also. It is your obligation to fulfill your contract with the group and support the majority decision.

If this were not true, there would be no point in taking a vote. The question at issue each time a vote is taken is: "Shall the approval and support of this organization be given to this proposal, or shall it not?"

Once the approval of the organization is given by majority vote, your individual approval and support go with it.

Individuals and Majority Decision

At one county meeting a zealous but inexperienced member fought hard and fairly against a raise in dues. The raise passed by a vote of 82 to 81.

Later on in the meeting the treasurer announced that he would like to have the members see him at the close of the meeting and pay their dues under the approved raise.

The member who had led the opposition jumped up and burst out, "I'm not paying the increase! Farm Bureau is a free, voluntary organization. I voted against higher dues and I don't intend to pay them."

The presiding officer explained, kindly but firmly, that the decision of the majority, once made, becomes the decision of all. He pointed out that, up to the time of decision by vote, every member has the right to oppose and to seek support for his opposition. But once the decision has been made by majority vote, it applies to all.

He went on to explain further that it is the right of any minority to protest or oppose, but the right of decision and the right to carry forward the program of the organization to the majority.

The protesting member saw the reasonableness of this explanation and started to write out his check.

At a Thursday night meeting of a county Farm Bureau out West there was wild enthusiasm. The county had exceeded its membership quota by more than 100 per cent. Such a victory called for a celebration. One member jumped up and proposed, "I move we hold a picnic Saturday to celebrate."

Mrs. David Brown led the opposition to the picnic. "The time is too short to get ready," she pointed out, and "everybody is busy on a Saturday. Moreover it looks like rain." She put up a strenuous argument against the picnic. However, the motion carried by a vote of 7 to 4.

The chairman announced, "The motion is carried. Since the time is so short, I'll call for volunteers to bring food." The first one on her feet was Mrs. Brown. "Mr. Chairman," she sang out with enthusiasm, "now that we've decided to have a picnic Saturday, let's make it the best one ever! Put me down for ten quarts of potato salad. I'm going home right now and put the potatoes on to boil."

Mrs. Brown proved that she had learned the most important lesson which a member of a group needs to know and practice. It is one of the fundamental rules of the game in voluntary organizations. But remember, Jefferson warned that it is one of the last lessons to be learned because it is the hardest.

Do you fully understand your obligation toward a decision by the majority? Can you profit by Jefferson's warning?

Just as the loser in a tennis match dashes over to his opponent and congratulates him, so the losers in a vote offer, not their congratulations, but their cooperation, because the majority decision has now become their decision also.

Groups and Majority Decision

Tremendous strength is generated when individual members understand how to support the principle of decision by the majority. With understanding comes unified support for Farm Bureau policies, decided by a majority vote of the members, directly or through their elected delegates.

Each group within Farm Bureau should also have the same understanding of how to support the principle of decision by majority vote. Each committee, each board of directors, each group of voting delegates, each county, community and neighborhood group, needs the same understanding and conviction of the necessity to support the decisions of the majority, if Farm Bureau is to realize its potential strength.

Suppose that Deerfield County has a surplus of certain products. At the county annual meeting a majority of the members vote in favor of a recommendation for higher support prices for these products. Later the voting delegates of the state and of the national organization approve a policy calling for lower support prices for the same products because they are convinced that the over-all good of the state and the country demands them. If the members of Deerfield County Farm Bureau are mature and experienced, they will close ranks in support of the lower tariff policy, which was judged wise by the voting delegates of the member state Farm Bureaus. They will help to give it a fair trial.

Suppose that one state adopts a recommendation favoring higher immigration quotas. But the voting delegates at the national convention of the American Farm Bureau Federation decide by majority vote against this recommendation. Do the members of the member state Farm Bureau which has lost out raise a cry of "Unfair?" Do they grumble and predict terrible results? No! Members and leaders support the judgement of the majority of the voting delegates and their state puts its shoulder to the wheel.

Actually, through the representative process, that state and all other member states have negotiated a series of agreements based on majority decision. These agreements are stated as policies. They are American Farm Bureau Federation policies, to be sure, but more importantly, these agreements are also state Farm Bureau policies with regard to national issues. After all, who determines these policies? The answer is – a majority of the official voting delegates of the member state Farm Bureaus – who are elected for this purpose.

Likewise, the resolutions agreed upon by the voting delegates of American Farm Bureau Federation become the policies of county Farm Bureaus with regard to national issues. Again, who selected the official voting delegates from the respective states? Who adopted these policies? The answer is – the official voting delegates from the county Farm Bureaus – who were elected by majority vote of the Farm Bureau members in the counties.

Thus, through the representative process, each member of Farm Bureau votes for his convictions. Voluntarily, all individuals and all groups within Farm Bureau give united support to the decisions made by the majority of the members.

How is majority decision enforced?

The letter of any decision of the majority is enforceable by rule. The spirit of any decision of the majority is enforceable only through the understanding and self-discipline of Farm Bureau members and Farm Bureau groups.

There is no law and no rule which can force you to abide by the decision of the majority – in spirit. If the vote goes against you, no one can prevent you from going around predicting glumly, "You mark my words. This will lead to trouble." No one can stop you from confiding to your neighbor, as you drive home from the meeting, "Lets see them make this thing work! Watch me drag my feet!"

You accept the decision of the majority of your own free choice. You support it because you have the maturity and the understanding to know that any organization must speak with one voice – and that voice is the voice of the majority of its members.

Jefferson was right when he said that real understanding of the soundness of majority decision is difficult. It is not too difficult to agree in theory that you will support the decision of the majority even if it is against you. The real test of your understanding of the principle comes when one of your pet ideas is defeated, or when a plan which you oppose, carries. Can you and do you about-face, fall in behind the decision, and give it your support? Do you "put the potatoes on to boil?"

The self-discipline which enforces the decisions of the majority is possible because members realize that one fundamental objective of Farm Bureau is to develop unity of purpose and action. It is the kind of organizational and individual discipline which is practiced because Farm Bureau leaders and members want to – not because they have to. Farm Bureau members have profited by Jefferson's wisdom. They speak with one voice. Farm Bureau has come of age.

WHAT IF THE MAJORITY IS WRONG?

When the majority decides a question and your side loses, does this mean that you can never raise your voice again in support of your opinion? Not at all. But your opposition should be withheld until a later and proper time. In a voluntary organization, such as Farm Bureau, you are not in the same position as a member of "Her Majesty's Loyal Opposition." Governmental bodies are based on a two-party system, with one party opposing the other much of the time. In a voluntary organization there is no permanent division into opposing groups. Instead of an obligation to continue pointing out defects in an adopted plan, you have an obligation to support it unless it comes up again for decision.

Suppose that the majority decides against something which is a matter of principle with you. What should you do? If things don't go too well under the decision, should you sit on the sidelines and snipe, muttering, "I told you so," or "Remember what I said?" If you do, you will only make yourself unpopular and get nowhere.

If the majority makes a wrong decision, and it sometimes does, inevitably the matter will come up for reconsideration at a later date. If it involves a policy, it will come up for reappraisal next year. When it does come up, if you have gone along with the decision, you are in a position to say something like this, "I opposed this action originally. I've watched it in operation for the past year. I have helped in carrying it out. Now I am more than ever convinced that the decision is wrong." This statement, from one who has cooperated, is bound to be listened to with respect.

How can you guard against mistakes by the majority? How can you insure the wisest possible decisions? By keeping every member fully informed on Farm Bureau plans and policies.

Are there alternatives to majority decision?

There are only two alternatives to majority decision. The first is decision by a minority, and no organization should be willing to be controlled by a minority. The other alternative would be for each individual or small group within the organization to follow its own differing decision. This would lead to chaos.

One large national association of professional men forgot this lesson and had to relearn it the hard way. After prolonged debate, and by a close vote, they placed the association officially on record as favoring an important piece of legislation pending before Congress.

On Monday, at the congressional committee hearing, the official representatives of the association presented the views and the decision of the organization. On Tuesday, a group of those who opposed the decision appeared, stating that they were coming before the congressional committee as representatives of a substantial minority of the association, and that they opposed the measure.

After the second group had been heard, and had left, the chairman of the hearing remarked to his committee members:

"Apparently this group can't agree. If its own members can't make up their minds, take a vote and stick together, we can't waste any more time with them."

Members of the Nebraska Farm Bureau House of Delegates vote on proposed policy during the General Session at their State Convention.

Taking Control from the Majority

Often people, who do not understand the sacredness of decision by majority vote, advocate decision by a higher vote, such as a two-thirds vote or a three-fourths vote, or even a unanimous vote. They mistakenly believe that a group is being cautious and conservative if it requires a high vote, and is protecting itself against hasty or unwise action. What they fail to understand is that, whenever a higher vote than a majority is required, the control is taken away from the majority and given to the minority.

When a unanimous vote is required on any proposition, control is taken from the majority and given to a minority of one member. This usually means that most action is blocked, because unanimous decision is seldom possible.

If a two-thirds vote is required to pass a measure, you can see that the one-third-plus-one minority group can prevent the passage of a motion which is favored by two-thirds-minus-one of the members – which is more than a majority. The one-third minority has the decision making power and the control. In other words, the members of the one-third-plus-one minority group have, in reality, two votes for every one vote belonging to the members of the larger or majority group.

Another way of expressing this is to say that if a motion requires a two-thirds vote, it takes two votes of the majority favoring it to equal one of the minority opposing it. Thus, control is taken from the majority and given to the minority of one-third.

Voluntary organizations protect vigorously the right of minorities to be heard, but support the right of the majority to determine the decision of the whole organization and the right to carry out that decision.

In using the words "majority" and "minority," it is important to realize that these are not permanent groups like Holsteins and Guernseys, or like peaches and pears. You find yourself in a minority on one question and perhaps two minutes later, on the next question, you are a member of the majority. The line-up may shift with each new motion. For this reason it is important to realize the rights and obligations of the minority and the rights and obligations of the majority.

If every member and every group of Farm Bureau understand the real meaning and obligations of majority decision and live up to them in letter and in spirit, Farm Bureau will continue to work in harmony and with unity of purpose of action.

Procedure in Meetings

CHAPTER 5

A state resolutions committee meeting of the New York Farm Bureau.

5

PROCEDURES PROTECT YOUR RIGHTS

Self-government is a simple concept. Lincoln expressed it as "government of the people, by the people and for the people." Unfortunately even the simplest ideas of government do not work unless they have suitable procedures which uphold them and unless these procedures are strictly observed. The procedures of our courts, our legislative bodies and of all the institutions of our country must be carefully adhered to if self-government is to be safeguarded.

For example, one of the fundamental concepts of self-government is voting by secret ballot. When you go to the polls at the next election you would be horrified if one of the election officials were to order you to vote for a particular candidate. You would protest violently if one of the judges were to look over your shoulder as you marked your ballot, and tell you how to vote. The right to vote by secret ballot is preserved and made possible only by the observance of a considerable number of detailed procedures. Many countries have a so-called "secret ballot," but not all of them have procedures which support a "secret ballot."

Few citizens understand how elaborate and extensive are the procedures set up in this country to protect a citizen's right to vote by secret ballot. The protective procedures begin at the polling place and extend clear through the higher courts. Without proper procedures, the right to vote by secret ballot is a mockery. Without proper procedures, no rights can be upheld.

We must observe proper procedures in everyday life – in mailing a letter, in buying and owning property, in driving a car, in shipping and in marketing produce.

A justice of the Supreme Court of the United States has well said, "The history of freedom in this country is the history of the observance of procedures." Suitable procedures, properly observed are the guarantees to individuals, organizations and countries of justice, liberty and freedom.

"The right of assembly" and "freedom of speech" are both established in the first amendment to the Constitution. These two rights are safeguarded by procedures which uphold, not only the rights and privileges of the group as a whole, but also the rights and privileges of each individual member. When members meet to propose ideas, discuss them and make decisions, their rights are safeguarded by a set of rules called parliamentary procedure.

Parliamentary procedure is a tiny section of the common law. Its rules are minimal for organizations. We all learn and observe traffic law. We all need to learn and observe the law of meetings. If you know the rules, and use them wisely, you can make a real contribution to meetings.

PRINCIPLES OF PARLIAMENTARY PROCEDURE

Every meeting which transacts business must, according to the law, observe the important principles of parliamentary procedure. Parliamentary procedure is an antiquated name for a simple, common-sense set of principles and rules which govern the conduct of meetings and which protect the rights of both the assembly and the individual. The rules of procedure are based upon a few principles which are easy to understand and easy to use.

In any meeting, only enough parliamentary procedure need be used to keep that meeting moving ahead in a businesslike and legal manner. Many parliamentary rules are needed only in unusual or complicated situations.

We all like to have a large vocabulary so that we will always have the right word for the particular occasion. However, we don't try to use all the words all the time. Just so with parliamentary rules – we need to know them all so that we can pick out the one that is needed in a particular situation. But most of the time we use only a few of them.

Parliamentary rules and principles are simple to learn and easy to apply, If you were appointed on a committee to draw up a set of rules for conducting business meetings, and had never read a book on parliamentary procedure, you would probably come up with much the same rules as those which form our present code.

Each rule is based upon a principle, and there are only a few fundamental principles which determine all the rules. Usually, if you do not know the answer in a particular situation, you can figure it out from a principle. Farmers and ranchers, like lawyers, naturally understand the use of principles. They work with them and know how to apply a general principle to specific situations. They know that there is a general principle that planting should be done when conditions are favorable. When the usual time for planting arrives, if the ground is frozen they do not plant – they apply the general principle and wait for favorable conditions.

These are the fundamental principles of parliamentary procedure:

1. **Parliamentary rules exist to help transact business and to promote cooperation and harmony.** If you see a person in any meeting who is using parliamentary procedure to confuse people, to trip them up, to win his way by trickery, he is misusing parliamentary procedure. He is acting contrary to its most fundamental principle. He is using a constructive tool in a destructive way. A hammer can be used to build a barn. It can also be used to hit someone. Technicalities should not be used to slip something over on uninformed members.

    Rules of procedure are developed to make it easier for each member to have his ideas considered, and easier for the majority to get things done with harmony and dispatch.

2. **The vote of the majority decides.** Self-governing countries and self-governing organizations are founded on decision by vote of the majority of the members. No group should tolerate control by a minority.

When a person joins Farm Bureau or any other organization, the common law of our country says that there is an understanding that he will abide by the decisions of the majority of the group.

True unanimous decisions are seldom possible. Most of the great decisions of our country – to declare war, to form a republic, to ratify our Constitution – were made by majority vote, and not by unanimous decision. Congress and our Supreme Court decide by majority vote, questions which influence the lives of all citizens.

3. **All members have equal rights, privileges and obligations.** It is the right of every member to be protected in the exercise of his rights and privileges. Since each member of a group has rights and privileges equal to those of every other member, it follows that he also has equal duties and obligations. It is therefore the duty of every member to carry his part of the work load and to accept his share of responsibility.

4. **The rights of minorities must be protected.** Self-governing bodies always protect the rights of minorities. Minorities and majorities are not permanent groups. They change as rapidly as sheep shift from small groups to a large group, and later back to small groups. If you are with the majority on an issue, it is important that you protect the rights of the minority, for soon you yourself may want the same protection.

5. **Full and free discussion of every proposition presented for decision is an established right.** This is the right of free speech, and every member has the right to "have his say," to "speak his mind." Free discussion

includes the right to question, to oppose, or to support whatever proposal is being considered.

6. **Motions have a definite order or rank in which they must be proposed and considered.** Only one main question can be considered at a time. Other motions can be presented and considered only in their proper order. This order or rank is as fixed as is rank in the army. Each motion must take its proper turn.

7. **Every member has the right to know at all times what the question before the assembly is, and what its effect will be.** No one should be asked to vote upon a motion unless he understands it. He always has the right to ask for further information. He can exercise this right until the chairman has announced the final result of the vote.

8. **Those to whom power is delegated must be chosen directly by the members or by their elected representatives.** If important power is to be given to officers, to committees, or to representatives or delegates, they should be chosen directly by the members or selected by whatever other method the members decide is best.

9. **The presiding officer should be impartial.** When presiding, the chairman is in somewhat the same position as an umpire or a judge – not a boss or a driver. A presiding officer must lead, but he must be both impersonal and impartial. This does not preclude him from answering questions by stating facts, or from making certain that the members understand the meaning and implications of the question before them.

These nine principles will guide you in any meeting – anywhere. They would help you to understand and to use the rules which are based on them and to make the correct decisions in the many situations not covered by rules.

PROCEDURE IN MEETINGS   page 65

(Photo courtesy Delaware Farm Bureau.)

Delaware Farm Bureau formed a committee to set up special events in recognition of their 50th anniversary. Committee members pour through old photographs to determine which will be used in a 50th anniversary history book.

(Photo courtesy Kentucky Farm Bureau.)

Kentucky Farm Bureau Legislative Committee members Wilbert H. Earley and Earnie B. Prewitt discuss farm issues.

Choosing a Parliamentary Authority

It is important for every organization to adopt a parliamentary guide as an authority which members can consult whenever their own rules do not cover a situation. The qualities to be considered in selecting a parliamentary guide are:

1. Completeness
2. Conformity to law
3. Up-to-dateness
4. Clearness

The most important question to consider in choosing a parliamentary authority is "Does this book conform to the law?" Many books on parliamentary law are written without knowledge of the law. These volumes lead organizations into litigation because they are at variance with the law.

When an organization adopts a book as its parliamentary authority, the book does not change any of the rules of the organization. It applies only to questions not covered by the rules of the organization. The order in which the law says that organizations must observe rules, is as follows:

1. The law of the nation and state
2. The charter of the organization, if it has one
3. The constitution and bylaws of the organization
4. Any special rules adopted by the organization
5. The parliamentary authority adopted by the organization

The rules contained in a parliamentary manual adopted by an organization apply only where there is no other rule of higher rank which has been adopted by the group. Any organization may adopt at any time a rule which differs from its parliamentary authority.

However, it is extremely important to adopt some parliamentary authority: first, because the courts presume that every organization is operating under some parliamentary authority and, in case of dispute, judge the organization by the rules of that authority; and second, because there must be a uniform set of rules to consult when there is no rule of the organization governing a situation.

Business Meetings

Chapter 6

(Photo courtesy New York Farm Bureau.)

New York Farm Bureau members debate an issues at their county annual meeting.

WHY MEETINGS ARE IMPORTANT

Meetings generate the motive power of Farm Bureau. In business meetings plans are made, recommendations are formulated, ideas are talked over, facts are presented, decisions are reached, members become better acquainted, and each meeting knits a stronger bond. Everybody goes home from a successful business meeting feeling that he has learned, made decisions, achieved results and enjoyed a pleasant fellowship – that by his presence he has contributed to his organization – that he has given and that he has gained.

Meetings small and large are the powerhouses of Farm Bureau groups. There is no such thing as a good community Farm Bureau with consistently poor meetings, or a poor Farm Bureau with consistently good meetings. A successful state Farm Bureau has worthwhile conventions. It would soon cease to be successful if its conventions were dull or poorly planned.

If you are really interested in Farm Bureau you will go to meetings – you will encourage others to go, too. You will do your part to ensure meetings worthy of Farm Bureau, its members and its aims.

THE MEETING PLACE

A suitable meeting place contributes to a good meeting. A courtroom with its overtones of formality and dispute, a church with its atmosphere of reverence, or a lodge hall pervaded by ritual may not be the best place for a meeting which should be friendly and encourage the members to express their opinions.

Twenty-five members, meeting in a hall built for a thousand people, are depressed by dreary space. Forty members gathered in a room designed to hold twenty persons are equally uncomfortable.

The meeting place should be suitable – with good lighting, heating, ventilation and acoustics. Comfortable chairs, arranged so that every member can hear well and see well, help immensely.

The aim in arranging a meeting place is suitability and comfort. A satisfactory meeting place does not necessarily require a large expenditure of money. The two outstanding meeting places in one state are: first, the impressive county Farm Bureau building in a large county, with air conditioning, soundproof walls and every up-to-date convenience, and second, in a neighboring county, the remodeled red schoolhouse, planned and done over by the members themselves, with all its traditional charm preserved – even to the red paint and the bell in the tower – but with modern kitchen, board room, meeting hall and office. Neither county would trade its meeting place for the other and both are equally suitable for excellent meetings. The reason is that both provide suitability, comfort and charm.

Preparations for Meetings

Most meetings devote a portion of the time to the transaction of business, another portion to a program, and usually some time to recreation and sociability. Each part must be planned. Good meetings do not just happen any more than good crops; they are the result of careful planning and intelligent carrying out of the plans.

Each meeting has a relationship to other meetings. No successful group waits until the end of one meeting and than asks, "Now what kind of a meeting shall we have next time?" They plan the program for the year so that there will be no haphazard drifting from one idea to another – no last minute makeshifts or failures.

Meeting planning divides into two main parts, both of which need careful, farsighted preparation: planning the business meeting and planning the program. This chapter discusses the planning of business meetings. Chapters 14 and 15 explain program planning for county meetings specifically, but the principles demonstrated can be applied to any Farm Bureau meeting or convention.

Diana and Jim Hanson listen as David Voldberg, presides at the Sweet Grass County Farm Bureau annual meeting in Montana.

Business meetings must be legal

Business meetings, in order to be legal, must follow an order of business and must be conducted according to the principles of parliamentary procedure. Parliamentary procedure is logical; it is common sense crystallized into rules of law.

Its main purpose is to help individuals and groups achieve what they want to get done in meetings. It is intended to make business meetings simple, clear and easy to conduct, and to enable people to work together effectively. The essential part of parliamentary procedure is its principles – the exact formal phrases are of less importance. It is the intent and good faith of the group that is vital.

For example, the technically correct way to ask a question is to say, "Mr. Chairman, I rise to a parliamentary inquiry," but if some member, who is unfamiliar with meeting procedures rises and says, "Mr. Chairman, I'd like to ask a question," the presiding officer should not rule him out of order, but should courteously ask him to state his question, and then answer it or call on someone else to answer it. The principle involved is that every member has a right to understand what is being discussed, regardless of whether or not he knows the correct form for asking.

The Pattern of Meetings

Business meetings of Farm Bureau and other voluntary organizations follow a pattern which is called the order of business. The bylaws of Farm Bureau generally list the order of business as follows:

1. Call to order
2. Reading or disposition of minutes of previous meeting
3. Reports of officers
4. Reports of standing committees or delegates
5. Reports of special committees
6. Unfinished business
7. New business, including important communications
8. Informal period (if desired)
9. Announcements, including introduction of new members and guests
10. Adjournment

The order of business is flexible. If a standing committee is not ready to report, some other committee is called on. For example, if the membership committee is not prepared to submit its report when called on, the presiding officer may ask the chairman of another committee, such as the policy development committee to report.

The business portion of a Farm Bureau meeting is kept separate from the program or informal part of the meeting, and the business meeting should adjourn before the program begins. Whenever any business is transacted, the chairman, if he is present, should preside.

Call to Order

Meetings should be called to order promptly at the appointed time, even though a quorum is not yet present.

A quorum is that number of persons who must be present in order to transact business legally. The number or percentage making up a quorum is stated in the bylaws. If a quorum is not present, don't waste the time until the required number come. Bring up matters of business and discuss them. When enough members have arrived to make a quorum, read the conclusions reached and ask if the members wish to vote to approve them or to discuss them further.

The presiding officer calls the meeting to order by tapping with his gavel and announcing for example, "The meeting of the Sutter Creek Farm Bureau will please come to order," or "The Twenty-fifth Annual Convention of the New York State Farm Bureau Federation is now convened."

Most meetings have a salute to the flag, a prayer, or a roll call at this point.

Reading and Disposition of Minutes

The presiding officer then calls on the secretary to read the minutes of the last meeting, or any minutes which remain unread. At the conclusion of the reading he asks if there are corrections to them. If there is disagreement on a correction, a vote must be taken. If a member moves to postpone the reading of the minutes, the chairman may say: "If there is no objection, we will postpone the reading of the minutes of the last meeting until the next meeting." If there is an objection, a vote must be taken on whether or not the reading shall be postponed.

Reports of Officers

At many meetings, the treasurer is the only officer called upon to make a report. Her report, unless it is an annual one, can be brief, since she reports in detail to the board of directors at frequent intervals. It lists important receipts and expenditures since the last meeting, giving the totals and the balance on hand. Members should have an opportunity to ask questions of the treasurer or to comment on expenditures if they wish.

If the treasurer's report is a final one for the year, it should be audited before the treasurer presents it to the meeting. A treasurer's report is never accepted until it has been audited. Routine reports are filed.

Efficient control over expenditures requires that there be authorization to spend money instead of waiting until a debt has been incurred and a bill presented. If an officer or member has been authorized to incur an obligation, it must be paid whether the organization approved the expenditure or not. Voting to pay bills already authorized and incurred is a formality without meaning.

In some Farm Bureaus other officers are given responsibilities which should be reported on at meetings from time to time.

Reports of Committees

The presiding officer next calls on the chairman of each standing committee who has a report, and then on the chairmen of special committees that are ready to report.

If the presiding officer is also chairman of the executive committee or board, it is often good practice for him to select some other member of the committee to give this report. Routine or progress reports are not voted upon. The presiding officer merely announces that they will be filed.

Unfinished Business

The presiding officer then asks for unfinished business. By finishing any uncompleted business before going on to consider new business, the meeting makes orderly progress. Unfinished business can refer only to:

1. A motion which was being considered and was interrupted when the last meeting adjourned
2. Any motion which may have been postponed at some previous meeting to this particular meeting

Unless some member brings it up first, it is the duty of the chairman to present any unfinished business.

New Business

The presiding officer then asks for new business. This is the members' opportunity to propose any motions which they wish to have considered and acted upon by Farm Bureau. If the presiding officer knows of matters which ought to be considered and which are not brought up, he may ask if someone wishes to offer a motion on the particular matter, or if the meeting wishes to discuss some specific problem informally.

Recommendations of committees and communications requiring action are considered under new business.

Informal Period

Some Farm Bureaus include an item in their order of business which they call the "Informal Period." During this period members may speak, ask questions, present ideas, or introduce new members, but they cannot propose any motions. Some member may wish to tell the assembly of a member who is ill and in need of help; another member may wish "to get something off his chest" by explaining it; others may want to ask questions or explain a problem which troubles them.

Announcements

The chairman calls on those members who she knows wish to make announcements; then she makes any of her own and asks if other members have announcements.

Adjournment

The motion to adjourn must always be voted upon, and if it carries, the chair announces that the meeting is adjourned. If no member proposes adjournment, the chair may ask if some member wishes to move to adjourn. If an hour for adjourning a meeting has been set, either in an agenda or program, or by the adoption of a motion setting a time, no motion to adjourn is necessary when the hour arrives. The chair announces the fact and declares the meeting adjourned. Also if there is no further business to be conducted the chair can ask, "Is there any further business?"

If there is no response, the chair can declare the meeting adjourned.

Agenda

An agenda is a list of proposals to be considered at a meeting or a convention. Agendas for meetings are usually prepared by the president and secretary. An agenda for an important meeting or convention is frequently prepared by the board of directors.

Sometimes agendas are adopted by vote of the members, but usually they are followed as a suggested program of proposals to be considered by the meeting. Proposals not included in the agenda may also be brought up at any meeting except a special meeting. Here is the agenda of one community Farm Bureau meeting:

Agenda for Meeting

MELROSE COMMUNITY FARM BUREAU – APRIL 9

1. Minutes of March meeting.
2. Treasurer's report.
3. Recommendations of Committee for new bridge over Elm Creek.
4. Recommendations of Committee on new zoning survey Midland area.
5. Election of new Farm Bureau representative to Community Council.
6. Decision on Farm Bureau's reply to proposal for a county library.
7. Decision on whether to change the planned program for the May meeting in order to invite the new superintendent of schools to speak.

The agenda for a board of director's meeting demands careful preparation so that there is ample time to discuss important problems thoroughly. It is easy to overcrowd and confuse the meeting by including correspondence and administrative problems which should be decided by officers, committees, or staff members. These problems, which are not the responsibility of the board, steal time needed to consider important policies which are the responsibility of the directors. If all decisions which are not policy problems of the board of directors are carefully pruned out of the agenda, the board will be able to concentrate on its important work.

Chairman's Memorandum for Guiding Meeting

The chairman needs a list of details which should be covered during the meeting. This is a personal memorandum and not an agenda. Agenda is the term applied to an agreed-upon and accepted list of proposals to come up at a meeting or at a convention. The memorandum supplements both the order of business and the agenda and, in addition, lists comments, suggestions and other details as reminders for the presiding officer. A memorandum needs no official acceptance from the group but is an indispensable aid to good presiding.

This memorandum is usually prepared jointly by the presiding officer and the secretary. If the chairman has a good memorandum, he will be able to conduct the meeting efficiently and keep it moving along smoothly. The following is an example of a well-prepared memo to guide the presiding officer:

Chairman's Memo
FRUITVALE FARM BUREAU – SEPTEMBER 3

CALL MEETING TO ORDER PROMPTLY AT 8:00.

1. Remind members meetings start promptly in order to close on time.
2. Appoint Ted Johnson temporary secretary since Secretary Roy Mason is in Memphis.

READING OF MINUTES

1. Ask Ted Johnson to read minutes of August 7.
2. Point out error giving correct date of annual meeting as October 1 instead of October 8.

REPORTS

1. Call for Treasurer's report-John Bailey. Comments on it.
2. Ask Vice-President Leon Frates, or Secretary Arlene Ellis, to report on meeting of Executive Committee.
3. Ask Louis Reeve to report on meeting of county Board of Directors.
4. Call on county Secretary Barnes to tell plans for county Farm Bureau annual meeting.
5. Ask Roy Peterson to give current membership figures.
6. Ask Ed Austin to report for committee on repairs to community hall driveway. Need motion authorizing expenditure of $116.00 for work.
7. Ask Ray Phillips, Chairman of county Policy Development Committee, to give outline on work planned.
8. Call for report of Program Committee – Janet Fulson, taking the place of Chairman Henry Phillips.
9. Ask Frank Cavendish to report for Extension Committee.
10. Call on Vera Sage to report on what neighborhood groups are planning.

UNFINISHED BUSINESS

1. Remind members that last meeting adjourned in middle of discussion on motion to send Executive Committee and officers to Clyde County to check the cost of their new Farm Bureau office building. Ask secretary to read motion. Call for further discussion before vote.

BUSINESS MEETINGS   page 79

NEW BUSINESS

1. Ask Secretary to read communication from state Farm Bureau Legislative Chairman concerning hearings on Senate Bill 3450.
2. Ask Secretary to read letter from Chamber of Commerce requesting cooperation in decorating Main Street for Rodeo Week.
3. If no motion offered on appropriation needed for Talent Find Team to go to state convention, ask if someone wishes to propose a motion.
4. Call for further new business.

INFORMAL PERIOD

Ask Sean to tell illness in Beatty family.

Ask Harold Godfrey to introduce new members – Fred and Leila Elliot.

Introduce Mr. Leland Chase, new county commissioner.

ANNOUNCEMENTS

1. Call on county Secretary Barnes to announce meeting of county Commodity Advisory Committee.
2. Ask Peter Haines to announce hearing on county tax assessments.
3. Call attention to next meeting: potluck dinner at 7:00PM, Ethel Franklin and Joe Davis in charge.

ADJOURNMENT – announce five-minute intermission before program.

Officers

Chapter 7

Scott Birch, Vice President of Orleans County Farm Bureau, John Blaisdell, President of Orange County Farm Bureau, and Charlie Huizenga, Delegate from Addison County Farm Bureau, discuss dairy issues and policies at the annual meeting of the Vermont Farm Bureau.

(Photo courtesy of Vermont Farm Bureau.)

The President or Chairman

A Farm Bureau president or chairman, whether he or she is the chief executive officer of a community, a county, a state Farm Bureau, or of the American Farm Bureau Federation, has two roles – each of which requires different abilities.

He is first of all a **leader**. As a leader he advances new ideas for projects. He expresses his ideas frankly and seeks to convert others to his beliefs. He is aggressive in getting new members and tactful in promoting understanding and unity among members. He seeks new ways to improve the Farm Bureau program – to build up the prestige and power of Farm Bureau. He weighs and studies policies.

She strives to unify and spark her group and to enlarge its accomplishments. She represents Farm Bureau to other organizations. The possibilities of what a president may do as a leader are limitless, and her role is as dynamic and many-sided as she chooses to make it.

The president or chairman, as a **presiding officer**, has a more definite and restricted role. He guides the meeting so that it moves rapidly and progressively forward. He is responsive to the wishes of the meeting, yet holds it firmly to its course. He refuses to allow members to confuse by technicalities or to wander from the subject under discussion.

She protects a member who has the floor in his right to speak without improper interruption, the minority in its right to be heard, and the majority from dilatory tactics. She is impersonal and impartial. She treats friends and opponents with the same courtesy and firmness.

He does not impose his opinions upon the meeting and does not usually speak on controversial questions. If, however, he feels that there are facts which he knows, and which the meeting should have, he states them from the chair. He retains his full voting rights.

An efficient chairman maintains order at all times. She is alert to the first signs of disturbance and acts promptly to restore order and attention. She uses her gavel sparingly. If a group becomes disorderly, a wise chairman refuses to entertain any business or to recognize speakers until order is restored. This will usually quiet the group.

He can do all this tactfully and not dictatorially. The chairman who is sensitive to the moods of the meeting knows when to encourage more discussion, when to hold it to the minimum, how to deal with embarrassing situations, how to encourage the shy member, and how to discourage the one who is overly fond of hearing his own voice. By his own conduct he exemplifies firmness and impartiality, courtesy and consideration.

She knows parliamentary procedure and how to use it.

Duties of the President or Chairman

1. To call the meetings to order at the appointed time.
2. To preside at all business meetings of the organization and of the board of directors.
3. To assign the floor to members who request recognition.
4. To state all proposals which are properly presented.
5. To restate all motions which are not clear.
6. To answer, or secure the answer to, queries on parliamentary points arising out of business before the meeting.
7. To restrict discussion to the question.
8. To put questions to vote and to announce the result of the vote.
9. To appoint committees as directed.
10. To act as the representative of the organization to outside persons and groups.

11. To perform all duties assigned him by the bylaws, or by vote of the board of directors or members.

Duties of the Vice-President

The vice-president's or vice-chairman's chief duty is to be prepared, if the president resigns, is away, or is incapacitated, to assume the office of president. When acting as president, the vice-president has all the powers, duties and privileges of that office.

Usually, he is assigned additional responsibilities by the bylaws; in Farm Bureau he frequently serves as chairman of an important committee, such as the resolutions or policy development committee.

The Secretary

The secretary is the chief recording and corresponding officer of the organization. He has all the rights of any other member and may propose motions, discuss them and vote. In Farm Bureau his chief duties usually are:

1. To provide the presiding officer, before each meeting, with a chairman's memorandum of matters which he knows ought to come up under each section of the order of business. This memorandum is usually prepared by the chairman and by the secretary working together.
2. To prepare a roll of members and call it when necessary.
3. To keep a careful and authentic record of business meetings.
4. To furnish the presiding officer or any member with the exact wording of any motion whenever requested.
5. To answer questions by reference to the minutes.
6. To read any papers or correspondence required by the assembly.

7. To authenticate all records by his signature.
8. To preserve all records, reports and documents of the organization which are not assigned to others.
9. To provide the chairman of each committee with a list of committee members, with instructions to the committee, and with any other material which will help it in its work.
10. To bring to each meeting a copy of the bylaws, parliamentary authority, policies of the group, and a list of all committees.

The Treasurer

The treasurer is the official custodian of the funds of the organization. She collects all money due and is responsible for its safekeeping. As disbursing officer, she issues checks when properly authorized, and reviews all vouchers and bills. She keeps an accurate record of all funds collected and disbursed. Her records are audited by a committee or by an accountant selected by the membership. She gives a detailed report to the board of directors at each regular meeting and a summarized report at each regular meeting of the organization.

He assists any officer or committee which needs information on the organization's finances. He usually serves as an ex officio member of the budget committee. This committee has the duty of preparing, in advance, a realistic estimate of income and expenditures for the next year. This estimate is the budget. Its provisions are not binding on the organization but are a guide for financial decisions. Even though a budget has been adopted, some items are necessarily exceeded and other items remain unexpended.

In Farm Bureau groups, the treasurer is usually bonded. If no provision is made for bonding, the treasurer should request, for his own protection and for the good of the organization, that a bond be provided.

The Past President

When a Farm Bureau president gives up his office, he may be tired and want a year's rest. But after that rest he is too valuable to shelve by labeling him "past president" and putting him back on the board of directors. He has served his full quota on the board and his job as a board member is done. The board needs new blood and a chance to develop the ideas of its members. Its members, like crops, should be rotated.

The former president, who has served well, needs a new job which is challenging and worthy of her ability and experience. As chairman of an important committee, as head of a new project, or as representative in a larger Farm Bureau group, she will face new problems which she is equipped to solve because of her valuable background or experience. **Farm Bureau needs her as an active leader of today and of tomorrow – not as a leader of yesterday.**

Directors

Directors, whether they serve a community, county, state, or the American Farm Bureau Federation, have similar responsibilities. As the directly chosen representatives of the members, a board of directors usually is given power to act for the organization whenever the organization is not "in convention assembled." Their power is a group power and is limited to actions taken as a board, unless the board delegates specific responsibilities to individual directors.

It is the duty of the board of directors to interpret adopted policies and to develop and set in motion a program to achieve policy objectives. Usually the board does this by requesting a committee, an officer, or staff members to formulate a proposed plan of action and present it to the board for consideration. The board then has the responsibility of approving these plans for carrying out policies and of checking later to see that they are followed.

Although the problem of administering a plan rests with the staff, with committees, or with officers, all of these are accountable to the board and report directly to it.

The board also exercises supervision over the finances and property of the organization. Directors consider, decide and approve the policies of administration which the organization follows. Once these policies have been determined, however, the duty and responsibility of actually carrying them out rests with the staff members or with others specifically assigned to do this work.

A director has no responsibility for the actual execution of the policy which he has helped to determine; administration is left to the officers, to the committees, or to the staff. It is their duty to execute the already determined policy.

Staff members do not give unasked-for advice to directors, officers, or committees and these members do not undertake or interfere with the work which belongs to the staff. Both spheres are clear-cut and entirely separate.

If a farmer plans to build a barn, he decides what type of structure he needs and then engages a contractor. The contractor does not undertake to tell the farmer what kind of a barn he should have, or impose his own ideas on his employer. The farmer tells the contractor his wishes as to size, price and use of the different portions of the building. When a contract has been signed, the contractor takes over.

The farmer checks at intervals to be sure that the work is being carried out according to specifications. He answers questions such as "Do you wish white or red paint?" He does not go out and tinker with the contractor's saw if it breaks down, or pitch in and nail the shingles on the roof. Being an experienced executive, the farmer, having once told the contractor what he wants done, leaves the actual work to him.

Said another way, policy interpretation and plans for policy execution are the responsibility of the board of directors. Actual administration is the responsibility and function of the officers, of committees and of the staff. Although all work closely together, each group has its own special field for which it is responsible. Since officers, directors, committees and staff members of Farm Bureau understand and recognize this division of responsibilities, groups function effectively and avoid confusion and misunderstanding.

Saginaw County, Michigan, Farm Bureau delegate Dale Kettler voices his opinion on an issue to the state delegate body.

(Photo courtesy Michigan Farm Bureau.)

Delegates

There are no American Farm Bureau Federation delegates as such. All are delegates of the states which selected them. Similarly, delegates to the state Farm Bureau are the elected representatives of the county Farm Bureaus. All delegates represent directly the members who chose them.

Farm Bureau follows the practice of sending uninstructed delegates to meetings and conventions. This practice is fundamentally sound and wise. Even though questions to be voted upon have been given wide circulation and publicity, there is always the possibility that amendments proposed at the convention may alter the wording, or that conditions may have changed.

It sometimes happens that, when delegates hear all of the facts presented from various sections of the state or county, they realize that their group might feel differently if the members at home could see the complete picture.

Every delegate should come to a meeting or convention with a general understanding of what his group desires, yet with leeway to vote as his intelligence and conscience direct him. Do not tie a delegate's hands, but depend rather on his good judgement and ability to interpret the wishes of those whom he represents.

If every delegate were given specific instructions on how to vote on every issue, there would be little point in holding a convention. The delegates could stay home and the instructions could be mailed in and summarized.

YOU ALWAYS REPRESENT FARM BUREAU

When a person becomes an officer, director, committee chairman, delegate or staff member of Farm Bureau, he no longer speaks only as an individual. His every statement carries with it the weight of his office. He becomes a part guardian of the official voice of Farm Bureau. In a sense he accepts a trusteeship to restrict his expressions of opinions, outside of Farm Bureau, to the decisions which have been made by Farm Bureau.

Officers, directors, delegates, committee chairmen and staff members of Farm Bureau need to be realistic. They must never forget that their acts and opinions as private citizens, and their acts and opinions as officers and representatives of Farm Bureau, cannot be separated in the mind of the public. For example, if an officer of Farm Bureau signs a petition, or announces his support of a controversial issue before Farm Bureau has determined its position, he is throwing the weight of his Farm Bureau office behind his action.

No matter how firmly he states that "I am not speaking or acting as a director of Farm Bureau, I am speaking only as a private citizen," those who hear his statement on the radio, or read it in the press, cannot, and will not, separate John Smith, private citizen, from John Smith, director of Oak Hill Farm Bureau. Their conclusion will be that, since John Smith is an officer of Farm Bureau and is for or against a proposition, Farm Bureau is for or against the proposition.

While a Farm Bureau member holds an office, he cannot divest himself of the influence which goes with that office by slipping off the coat he is wearing as an official representative of Farm Bureau and slipping on the coat of a private citizen.

Certainly the responsibility of holding office in Farm Bureau is not intended to limit the private acts of citizens. Public acts and expression of opinion on controversial issues, however, should await the official decision of Farm Bureau members. Every officer and member of Farm Bureau would do well to keep in mind his responsibility to exercise restraint in acting ahead of membership decisions on matters which concern Farm Bureau members and involve Farm Bureau policies.

To nonmembers and other organizations Farm Bureau speaks with one voice. That voice expresses the agreed-upon opinions of the many groups within Farm Bureau and is the result of a majority vote of its members.

Patience in waiting for Farm Bureau's decision may mean that the Farm Bureau officer or representative sacrifices some of his outside leadership opportunities. But within Farm Bureau his personal status and value increase, because Farm Bureau members respect his recognition of responsibility to the office which he holds and to the organization which he represents.

SAMPLE JOB DESCRIPTION
COUNTY FARM BUREAU BOARD OF DIRECTORS

FUNCTION

To control and govern the business of the county Farm Bureau.

REPORTABILITY

The board's authority is a collective one, and individual directors have no executive powers except when acting collectively as a board or when delegated specific authority by the board. In the collective decision making, individual directors must subordinate the special interests of their districts or committees to the best interest of the county Farm Bureau.

RESPONSIBILITY

1. The board shall determine the goals, objectives and administrative policies of the county Farm Bureau within the framework of the articles, bylaws and policies established by the membership.
2. The board of directors should become familiar with and adhere to the relationship agreement between the state and county Farm Bureau.
3. The board shall elect or appoint such officers and employees as authorized by the bylaws and determine any remuneration.
4. The board has the duty to protect and control the corporate assets and to maintain the solvency of the county Farm Bureau through the adoption of sound fiscal policies including an annual budget and audit.
5. The board shall delegate the administrative responsibilities to the county president or designated employee.

6. The board shall approve the selection and appointment of county committees to carry out county programs and activities.
7. The board shall determine, review, appraise and support the activities, programs and economic services offered to members by the county Farm Bureau.
8. The board shall see that a yearly membership drive to maintain a growing organization is planned and conducted.
9. The board shall require progress and financial reports from those to whom responsibilities have been delegated.
10. The board must maintain adequate minutes of all meetings since they will constitute legal evidence of actions taken by the board.
11. The board shall plan and conduct a county annual meeting at which:
    - Progress and financial reports are adopted,
    - Policies on county issues are adopted,
    - Recommendations on state and national issues are determined, and
    - The delegates, officers and board members are elected.

RELATIONSHIPS

To inform other county Farm Bureaus, the state Farm Bureau and affiliated companies of issues that may affect their well being and the board should project a positive image to the community and other agricultural organizations.

MINUTES

CHAPTER 8

McCone County, Montana, Farm Bureau Secretary Monica Switzer and Fred Thoney, County President, compare notes at their county annual meeting.

What Minutes Contain

Records of Farm Bureau meetings vary all the way from the brief handwritten page of minutes of a local meeting to the complete record of the stenotypist at the annual meeting of the American Farm Bureau Federation. The official minutes of any meeting or convention are those prepared and signed by the secretary and approved by the organization.

Minutes, in the opening paragraph, give the following facts:

1. Name of organization
2. Kind of meeting (regular, special, or adjourned)
3. Time, place and date
4. Presiding officer

Minutes are the record of motions introduced, actions taken, reports made and votes cast. Each motion and the name of the proposer is recorded, whether it was passed or lost. It is good practice to record also the name of the seconder of important motions.

No record of discussion is kept.

A vote by roll call or by ballot or a counted vote is recorded.

Written or printed reports of officers and committees are best kept in a separate book. A notation is made in the minutes that the report was given and may be found on a certain page in the report book.

Minutes contain no opinions or comments of the secretary. While these are often interesting and entertaining, they are out of place and frequently lead to disagreement and misunderstanding. Phrases such as "an outstanding speech," "a violent argument," or "a generous response of approval" should be omitted. The secretary has no more right to use comments and descriptive phrases in his minutes than the treasurer has to include them in his report.

If the treasurer should read an item thus, "Paid for printing of annual programs, which many consider a foolish waste of money, $27.50," you would be startled.

When members wish to record gratitude or appreciation, this is best done by a vote of thanks or a resolution of appreciation.

Approval of minutes

Minutes are signed by the secretary, and must be approved by the members at a meeting. After the minutes are read, the chair calls for corrections. If there are none, she announces, "The minutes are approved as read."

It is unwise to make a practice of repeatedly postponing the reading of minutes, because the longer the time that intervenes, the harder it is to make corrections.

Sample minutes

JONES CREEK COUNTY FARM BUREAU:

The regular meeting of the Jones Creek County Farm Bureau was called to order on May 15 at 8:10 p.m. at the Lower Jones Creek Schoolhouse, Chairman Thomas Harding presiding.

Twenty-nine members present.

The secretary, Mary Lynn Henes, read the minutes of the meeting held April 17. Andrew Carter called attention to a mistake in listing him on the 4-H County Fair Committee instead of his brother, Albert Carter. Hattie McKinley pointed out that there was an omission in recording the representatives of the Women's Committee who attended the District Meeting at Kirksville. Ruth Follet and Elizabeth Austin should be added. Both these corrections were made, and the minutes were approved.

The Treasurer gave an itemized report showing totals of:

    Receipts $ 649.50
    Expenditures   1,720.00
    Balance in bank   3,809.93
    (See page 221, Report Book)

Vice-Chairman Moffitt gave a brief report of the Executive Committee meeting held on May 10. He stated that the committee had approved a recommendation that Jones Creek Farm Bureau support the proposed changes in the county zoning ordinance.

George Hayes, Chairman of the Membership Committee, reported that most of the membership work for the year has been completed and that twelve new members have been added since the last meeting.

Pete Longacre reported for the Program Committee because Chairman Michael Rountree was unavoidably absent. He announced that State Delegate Thomas Goodall had been invited to the June 14 meeting to discuss pending legislation in our state legislature.

Under unfinished business, the Chairman announced that at the last meeting a recommendation on state license fees for trucks had been postponed until today's meeting. The Secretary reread this recommendation:

"It was moved by George Peterson, seconded by Robert Fowler, that the following recommendation be sent to the state Board of Directors: 'Jones Creek County Farm Bureau recommends that the state Farm Bureau vigorously oppose changes in the state license fees for trucks as provided in Senate Bill No. 6920.'"

The chair called for a vote on the motion which carried.

The Chairman then called for new business.

It was moved by Carl Erickson, seconded by Floyd Peterson, "That the Jones Creek County Farm Bureau support the proposed changes in the county zoning ordinance as recommended by its Executive Committee, and that a copy of this recommendation be sent to the county Board of Supervisors."

Carol Erickson moved for the adoption of the following statement: "That the Jones Creek County Farm Bureau support the proposed changes in the county zoning ordinance and that

MINUTES page 97

the county President send a letter of support to the County Board of Supervisors." The motion was seconded by Andre Carter. Motion carried.

It was moved by Peter Argall, seconded by Thelma Hayes, "That we authorize the expenditure of $175 for caterers to prepare the dinner to be served at the June meeting honoring State Director, Warren Goodall."

It was moved by Mary Thomas, seconded by Ethel Lorimer, that the motion be amended by adding that the meeting will be held at the Community Building and all former county board members be invited to attend." Amendment carried.

It was moved by Tom Atherton to further amend the motion by increasing the $175 to $275." The amendment was seconded and approved. Motion as amended carried.

The chairman asked if the members wished to hold the annual picnic in July or August.

It was moved by Trigg Rice, seconded by Henry Soter, "That we hold our annual picnic the third Saturday in July at the home of the Peter Starbuck family." A rising vote was called for. The vote was 18 in favor, 7 opposed. Motion carried.

There being no further business, the chairman called for announcements.

Membership Chairman George Hayes introduced new members who have recently joined Farm Bureau. These are: George and Lola Higgins, Tony and Anna Franzini and Arthur and Muriel White.

County Secretary Henes announced that a hearing would be held in the Community Room of the City Hall on May 21 at 8:00 p.m., on rate changes proposed by the telephone company. The meeting is open to all.

Peter Longacre announced that State Director Goodall would like to have a short session with the Legislative Committee at the close of the next meeting.

County Agricultural Agent Black announced that Dr. John Powers of the Extension Service would speak on May 27 at 8:00 p.m. at the Hillside County High School Auditorium, on Agriculture's Stake in Foreign Markets. He invited all members and their friends to attend.

Roger Thomas moved to adjourn. Motion carried. Meeting adjourned at 10:35 p.m.

     (signed) Mary Lynn Henes, Secretary

Discussion in Meetings

CHAPTER 9

Arkansas Farm Bureau President Andrew Whisenhunt speaks during the business session of the American Farm Bureau Federation annual meeting.

(Photo courtesy Arkansas Farm Bureau.)

Talking Things Over

James Madison once declared to the House of Representatives, "For my part, I had rather do less business and do it well, than decide measures before they are fully discussed and understood."

This is an assertion of the importance of free, open and full discussion – of the right of every member to talk things over and to understand the question on which he is to vote.

Farmers and ranchers believe in getting the facts on a question before making decisions. They know that freedom of discussion is necessary to uncover the truth. They face facts and base decisions upon them in their farm business. No farmer would buy a new piece of land without finding out the facts about the water supply or testing the soil. He wouldn't go "on a hunch." Farmers want the facts on every question before they vote on it – and that is not only their right but their obligation.

Most farmers are open-minded and glad to listen while the facts are presented. If a member were to go to a meeting with his mind closed to any suggestions or facts which might develop during discussion, he would be useless as a participant or a decision maker.

Principles of Discussion

One of the fundamental rights guaranteed to the citizens of our country is the right to assemble and to talk things over. The reason why a group of people can usually make a wiser decision on a problem than one person alone, is that the group talks over and listens to various opinions on the problem. Members who have heard differing viewpoints on a question are usually able to harmonize them into substantial agreement. Without discussion, or debate as it is called in meetings, there could not be intelligent group decision.

One important principle in talking things over is that the right belongs to each member equally. Mr. Jones, who is a polished orator through his experience in the state legislature, has no more right to discuss a motion or to speak longer or more often than Mr. Smith, who may be a poor speaker. Every member has an equal right to the time and attention of the assembly.

Another principle is that it is important that as many members as wish should take part in discussion. It is the duty of the presiding officer to guide and stimulate debate so that it will arouse the interest of the members. It is his job to create an atmosphere which is warm, inviting and favorable to free and frank discussion. It is also his responsibility to see that all members are given the opportunity to present their viewpoints.

A third principle is that the right to debate should usually not be cut off, even by a motion to vote immediately (previous question), until all important viewpoints have been expressed. Nor should a chairman hurry a vote and thus cut off the right of members to speak. A vote that is won by pressure, trickery, or unfairness is likely to be reversed later. It is far wiser to take the time to convince members on a subject than to try to "railroad" a motion through.

Macaulay said, "Men are never so likely to settle a question rightly as when they discuss it fully."

This does not mean that every member must take part in discussion. Many persons do not like to speak in public. Furthermore, when six or seven speakers have talked on a topic, there is often little left to say. No chairman should embarrass a member who has not spoken by calling on him to give his opinion, but should stress rather the importance of every member's listening and voting. It is a member's vote rather than his speech which is important. Discussion is a **privilege**. Voting is a **duty**.

If you can vote more intelligently by sitting still in a meeting and listening, then by all means be a listener. You are doing your full duty as a participating member if you listen, weigh arguments and vote.

The great basic principle of discussion is that a member may say anything he wants about a motion or a proposition, but he cannot say anything which reflects on the maker of the motion. He may attack Mr. Thompson's resolution as "ridiculous," "vicious," or "unfit for this group to consider," but he cannot reflect in any way on Mr. Thompson. In other words, discussion is concerned with **measures, not with men** – with **ideas, not personalities**. A real understanding and observance of this principle is the mark of a mature member and of a mature organization.

What is good discussion

Discussion is good when it throws light on all sides of a question and brings out the truth. It is good when it progresses steadily toward a decision.

Most discussion is friendly and helps to make the subject clear. When there is no difference of opinion it is only necessary that each person understand fully the question on which he is to vote. It is each member's privilege to contribute what he can toward reaching a wise decision.

The goal of agreement is sometimes reached, however, through the presentation of opposing opinions. One group holds one conviction and other groups champion different opinions. Vigorous spirited debate on a question is one way of bringing out the facts. Such a clash of opinion is basically friendly and often clears the air and helps make agreement possible.

Contributing to discussion

Making a speech or presenting arguments are not the only ways of taking part in discussion. Even more valuable than the ability to make a speech is the ability to:

DISCUSSION IN MEETINGS page 103

1. Analyze the discussion situation and the need at any given moment.

2. Contribute the idea, suggestion, or motion which will meet the situation.

A coach sits on the sidelines and watches a game, analyzes the situation, and determines the most effective plays. During discussion, experienced members seek to analyze the situation and decide what argument should be advanced, what questions asked, what motion proposed, to meet the situation at any particular moment.

Timing is one of the fundamental abilities which every member needs if he is to succeed in discussion. Timing means knowing when to use a suggestion, a question, or an argument so that it will be most effective.

Here are some situations which commonly arise during discussion and examples showing how they may be met effectively:

1. If you are puzzled by something that is said, or by something left unsaid, **ask a question**. Chances are a lot of other people are puzzled by the same thing and will be grateful to you for voicing their question.

One Farm Bureau meeting was discussing the motion that the expenses of delegates be paid "as in previous years." Many members were puzzled as to what the exact practice had been, yet hesitated to reveal their ignorance. One member asked, "What expenses **have** been paid in previous years?" and cleared up the whole matter for everyone.

2. If the meaning of some word or phrase or idea is being misunderstood, get up and **try to clarify it**. If something is clear to you, your explanation will help others to think clearly.

Recently an amendment was proposed to the bylaws of a fraternal organization. It read, "A majority vote is required to approve the purchase of any real property." Several members kept talking about the fact that their organization would never be able to purchase any property because it would be impossible to get a majority of the members out to a meeting.

One member rose and explained that a "majority vote," unless otherwise stated, meant a majority of the votes cast, and not a majority of the members of the organization. This brief explanation clarified the discussion instantly.

3. If you think that a change in the motion which is being discussed will make it more agreeable to most of the members, **propose an amendment** to the motion.

One group had discussed a proposal to purchase new office equipment for almost an hour. It was evident that the members agreed that new equipment was necessary. It was also evident, although unstated, that some of them feared the extravagance of the county secretary. One alert member sensed the difficulty. He promptly proposed, "I move that the motion be amended by adding the words, 'and that a committee, headed by the treasurer, be selected to work out estimates and submit them at the next meeting.'" This change in the original proposal brought immediate agreement.

Another way to alter a motion or recommendation is to refer it to a committee to reword.

4. If there is confusion because a motion is composed of several parts, each independent of the others, **ask for a division of the motion** so that each part may be decided separately.

Recently, a large convention was struggling with the motion, "that we hold our convention in Detroit next year, and in Miami the following year." Discussion became hopelessly tangled. Some wanted to go to Miami first.

Others felt that if they went to Detroit next year they ought to go out west the following year. Some members didn't want to go either to Detroit or Miami. Discussion wandered in a maze, instead of marching ahead.

Finally a young delegate stepped to the microphone. "We are getting nowhere because we are trying to discuss two motions at once," he stated. "I call for a division of this question into separate motions."

There was enthusiastic applause. The delegates quickly decided the location of each convention in turn. That delegate made the right contribution to discussion at the right moment.

5. If debate becomes too heated and members grow tense, try a humorous remark or a brief story. A bit of humor will often ease the tension and help to bring about agreement.

At one county Farm Bureau meeting, members had been discussing for an hour and a half, the question of building a new county Farm Bureau office building. The discussion had ceased to be objective. Members were not willing to listen to any compromise of their viewpoint.

Finally a farmer in the back row, who had not spoken, stood up. "Mr. Chairman," he drawled, with a friendly grin, "we're getting no place fast. We're acting like a centipede that Abe Lincoln told about. The centipede's name was Cyril. Cyril's legs all wanted to go in different directions at the same time. When Cyril made up his mind to amble east to a sunnier rock, often the thirty front legs would decide to go south, the forty rear legs north, and the center legs would mill around indecisively. In desperation, Cyril addressed the legs one day. 'Legs,' he said, 'all of you want to go in different directions, so I can't go any place. I just stand here and quiver.'"

Everybody laughed, and one member called out good-humoredly, "If Cyril solved it, so can we." In a few seconds, the member in the rear row broke the log jam of tension, and agreement soon followed.

Another way to relax meeting tension is to **propose a short recess**. Members return refreshed and in a more tolerant frame of mind. Frequently when discussion is strained, you can restore a more reasonable attitude merely by making a few remarks in a low and pleasant tone of voice.

Sometimes a motion to **refer the question to a committee** will give opportunity for a more considered decision, or a motion to **postpone the problem to the next meeting** will meet the need.

6. If a speaker is getting off the track, call the chairman's attention to the fact by **rising to a point of order**. Or rise to a parliamentary inquiry and **ask a question** which will bring the speaker back to the subject, such as "Isn't the point of this question whether or not Farm Bureau wishes to spend additional money for printing?"

7. If discussion goes on past the point where it is helpful, propose a motion to **limit debate**, or to **vote immediately**.

8. If there are several differing opinions, **point out the facts on which all agree**. Even one point of agreement can serve as a basis for further agreement.

A Farm Bureau meeting was trying to decide on a date for their annual picnic. Seven dates were proposed. Each date was opposed by a different group because they could not attend. Twenty minutes were wasted in discussion.

One member rose and said cheerfully, "We agree on two things, anyhow – one, that we want a picnic, and two, that it should be held in the summer. No date is agreeable to everybody, but we can pick the date which will suit the largest number of members. Let's vote on each of the dates, beginning with the earliest, and pick the one receiving the highest vote.

In three minutes the problem was settled.

9. If agreement seems hopeless, **propose a compromise** around which opinions may rally. Give a brief, practical review of the facts brought out by each side. Then offer the new compromise proposal.

A Farm Bureau Group was trying to pick a location for their county office. One group favored the Blake Building at the extreme north edge of town "Because it is lower in price, convenient to the shopping area, and a large number of our members live north of town." Another group stood firm for the Du Four Building at the south end of town "because it has much better parking facilities, is downstairs, and almost as many of our members live south of town as north." The two groups were deadlocked. Neither wanted to cross town.

Finally a member of the County Office Location Committee arose. "The Northerners claim a saving of $100 per month, conveniences of a second-floor location, and the shopping area. They don't want to drive across town to reach the Southern site.

"The Southerners maintain that a ground floor is better, and contend that their favored site has better parking. They don't want to drive across town to reach the Northern site.

"Now actually, a lot of us don't live either north or south of town, but in other areas of the county. It is evident that the north and the south aren't going to get together. A decision for either location will leave about half of our members unhappy. This is harvest time and we can't stay here all night. I propose this compromise:

"The top floor of the Penny Building costs only $25 more than the other two sites. It has a good parking lot and is right in the center of town. When all of you people are already driving into town it will be easy to meet in the center. I move that we authorize the committee to take a five-year lease on the top floor of the Penny Building."

These and many other ways of contributing to discussion situations require only a brief suggestion and not a speech. Their value depends upon their use at the proper time.

(Photo courtesy Iowa Farm Bureau.)
Voting delegates at an Iowa Farm Bureau Annual Meeting.

(Photo courtesy Illinois Farm Bureau.)
Daryl Brinkman and Mike Poettker, Clinton County Farm Bureau delegates discuss the merits of a proposed policy at the Illinois Farm Bureau annual meeting.

CHAIRMAN'S DUTIES DURING DISCUSSION

It is the duty of the presiding officer to see that the principles and rules of good discussion are observed; that discussion moves along and makes progress; and that the rights and privileges of all members are maintained.

It is also her duty to preserve order – to prevent whispering or walking about, heckling, or improper interruptions. When a member has been given the floor to speak, the presiding officer protects him from interruptions. While he is speaking, members may ask questions only if the speaker is willing to stop and answer them. Both questions and answers are directed to the chairman.

The chairman keeps the question under discussion clearly before the meeting and repeats it whenever necessary. It is his responsibility to see that discussion keeps to the point. A member may tell a story, read a short quotation, or make a comparison, if it has a direct relation to the subject which is being discussed. If a member departs from the question, the chairman interrupts him courteously but firmly and directs him to speak on the subject. If a member should forget his meeting manners, and become personal or rude, the presiding officer immediately interrupts him and directs him to conduct his discussion in an orderly and courteous manner.

It is also the duty of the presiding officer to see that no member talks too long. There is no fixed rule as to how long a member may speak, but any group may place a time limit on speeches. A long speech is seldom as effective as a short one. If a member takes too much time the chairman may tactfully interrupt him. "Pardon me, Mr. Brown; we have only a few more minutes before the adjournment. Please conclude your remarks quickly so that others may speak."

The presiding officer makes sure that opportunities for discussion are divided between those who favor a proposition and those who oppose it. If a member who favors a motion has just spoken, and several members rise and request recognition, the chairman may ask, "Does anyone wish to speak against the motion?" and then recognize someone who does.

No one admires a bossy chairman. Yet no one respects a chairman who does not lead, but appears to be pushed along by the members of the group. A presiding officer maintains firm control of the meeting at all times and guides, but does not direct, the group.

INFORMAL DISCUSSION

There is a general rule that usually there can be no discussion until a motion has been made and seconded and stated to the meeting. There are times, however, when discussion must precede a motion. To take care of such situations, the process of "informal discussion" has developed.

Suppose that the members of a Farm Bureau realize that their meetings are poorly attended because Monday night is inconvenient. A more convenient time must be found. Everyone knows this, but no one knows which night will be best. Therefore no one can propose a suitable motion. Such a situation can best be solved by informal discussion of the problem.

A member may request recognition and say, for example, "Mr. Chairman, we have to find a more convenient meeting time. All of us realize this. I move that we discuss this problem informally."

If this motion carries, the chairman calls for informal discussion on the problem of a more convenient night. During this informal discussion there is no limit to the length of the speeches, or the number of times a member may speak. Members have more leeway to bring up ideas which do not bear directly on the question under discussion. As soon as there seems to be general agreement, a member proposes a motion embodying the ideas which have developed from the informal discussion:

"Mr. Chairman, I move that hereafter we meet on Friday evenings at eight o'clock."

The proposal of this motion automatically terminates the informal discussion, and the regular rules of discussion again apply. Informal discussion is a useful method when a problem needs consideration before a suitable motion can be proposed.

Farm Bureau's Neighborhood and Community Discussions

Where do the most important Farm Bureau discussions take place? At the American Farm Bureau Federation convention? At the county annual meetings? These are important, but not any more important than the discussion that goes on in the small neighborhood and community groups.

In the evenings and on weekends, throughout the United States, thousands of small groups of Farm Bureau neighbors are discussing ideas for better agriculture and for better government. A dozen cars parked near the home of a Farm Bureau member may indicate that around the fireplace or

DISCUSSION IN MEETINGS

dining-room table, in the kitchen or on the porch, farm families are enjoying an evening of informal discussion.

Sometimes, there are only three families, and sometimes there are twelve. Sometimes the group is a tiny community Farm Bureau, sometimes a committee, and sometimes it is a gathering of neighbor members of Farm Bureau with kindred interests who have come together to talk things over. Sometimes families go the first time because they "don't want to disappoint John and Emily." They continue to go because they know the value of these small community and neighborhood groups.

They enjoy the apple pie and coffee, or the ice cream and cookies, or the doughnuts and milk. But they respond even more keenly to the lively arguments, the shared problems and the genuine achievements which come out of those discussions. Here in John Jones' living room are born and developed ideas which may later become Farm Bureau policies. Here also are exchanged hundreds of facts and knowledge helpful to every farmer. Frequently the group listens to a brief explanation of a problem. Then all of the members join in discussion. Often a short summary of the problem or of a fundamental Farm Bureau policy lends interest and substance to the discussion which follows.

Most of these groups do not meet haphazardly. They are part of a well organized and well serviced program.

Each of these little groups meets in a relaxed atmosphere of neighborliness, of freedom to talk informally, of sociability and of encouragement to try out new ideas. Each member gives, and each member receives.

It is impossible to estimate either the power or the value of these thousands of groups. They prepare for wider service in larger meetings, they educate and inspire, they save the time of larger meetings because the ideas introduced have been perfected in the small group, and they unite farm and ranch families in common cause.

Voting

Chapter 10

County delegates considering issues at the Maine Farm Bureau annual meeting.

Alchua County, Florida, Farm Bureau board of directors meet to discuss county activities.

The Right to Vote

A member's right to vote is the most fundamental of all his rights of membership. It is even more important than his right to speak. There are some members who do not like to speak in meetings, and there is no reason why they should be obliged to do so.

To **consider** whatever ideas are offered in a meeting, however, and to **vote** on them is not only a right but a duty which must be fulfilled. Sit quietly and weigh arguments if you wish, but if you do not vote, you are not fulfilling your obligation as a member.

Methods of Voting

Members may vote by:

1. Voice
2. Rising (or by raising hands)
3. Roll Call
4. Ballot
5. Unanimous consent

When considering any question, members have the right to decide how it shall be voted upon. During discussion any member may propose a motion determining the method of voting on the main question and this is decided first.

Most motions are decided by **voice** vote. The presiding officer says, "Those in favor of the motion (stating it) say 'Aye'...Those opposed, 'No.'"

If the chairman has no doubt about the result of the vote, he says, "The motion is carried," or "The motion is lost."

If he is uncertain as to the results of a voice vote, he should take it again. Any member who wishes to have a doubtful voice vote counted by a rising vote or by raising hands, may simply call out "Division!"

A vote by **rising** or by **raising hands** is often used when it is important to have an accurate count. Either method may be used to verify a voice vote when "Division" has been called for.

A vote by **roll call** is accurate but takes considerable time. It is ordinarily used chiefly when members are voting as representatives and delegates of other groups, so that there will be an official record of each member's vote.

Voting by **ballot** is used when secrecy is desirable. A vote by ballot is the only way in which a member can keep his opinion a secret.

Voting by **unanimous consent** is used only for routine or commonly agreed-upon questions. For example, a chairman might say, "The credentials committee is not ready to report, so, if there is no objection, the program committee will present its report now. Is there any objection?...There being none, the program committee will present its report." This is called "voting by unanimous consent."

Or a member may say, "The Mayor has arrived to speak to us and her time is limited. I ask unanimous consent to interrupt this discussion and hear her at this time." The chairman asks "Is there any objection?"...and if there is none, no vote is taken.

Voting by unanimous consent on routine motions is one way of saving time.

When a Member Cannot Vote

A member cannot vote upon a motion in which he has a direct personal or financial interest. For example, if Hazelwood County Farm Bureau is considering a motion to award a contract for earth moving to a member who has placed a bid, this member cannot vote upon the motion because he has a direct financial interest in it.

If, however, a considerable number of members have a financial interest in a motion, they may vote on it. For example, if the amount of money to be allotted to each of seven delegates attending a convention is being determined, the delegates would have a right to vote.

Changing Your Vote

A member may change his vote up until the time that the chairman has announced the final result. To do this he rises and, without waiting for recognition, says: "Mr. Chairman, I wish to change my vote from 'No' to 'Yes' (or from 'Yes' to 'No')."

A vote by ballot cannot be changed after the ballots have been collected or deposited.

Voting by the Chair

Normally the chair of a deliberating body does not cast a vote. However, according to Robert's Rules of Order, the chair does have every right to cast a vote.

"If the presiding officer is a member of the assembly, he can vote as any other member when the vote is by ballot. In all other cases the presiding officer, if a member of the assembly, can (but is not obligated to) vote whenever his vote will affect the result – that is, he can vote either to break or to cause a tie; or, in a case where a two-thirds vote is required, he can vote either to cause or to block the attainment of the necessary two-thirds."

Voting by Proxy

Voting by proxy is unsuitable in most organizations where members have equal rights. In corporations, where a member has as many votes as he has shares of stock, proxy voting is customary.

What Is a Majority Vote?

Majority vote means a majority of the legal votes cast, unless the term is qualified differently. When the term majority vote is used in the bylaws or in a motion, the phrase should be explained by stating what kind of majority is meant. For example:

1. A majority of the total membership
2. A majority of the members in good standing
3. A majority of a quorum
4. A majority of members present
5. A majority of the legal votes cast

How Many Votes Decide a Question

If the vote is called for on a motion and members are paying attention to something else, perhaps only the maker of the motion will vote "Aye." If no negative vote is cast, the motion is carried by a majority of one vote.

A tie vote does not carry a motion. An assembly may decide to resolve a tie vote by any means it chooses.

Don't Fail to Vote

Sometimes you hear members explain why they did not vote by saying: "I knew it was going to carry anyway," or "I didn't understand the question," or "I wasn't vitally affected," or "I didn't want people to know how I felt."

If a member does not understand a question, he always has the right to ask the chairman to explain it further. There is no excuse for failure to vote. The fact that you are not vitally affected, aren't anxious to reveal your opinion, or feel that the motion will carry or lose anyway, is no reason to shirk the most important duty of every member – **to vote**.

A Farm Bureau member presents his views at a county annual meeting in New York.

NOMINATIONS AND ELECTIONS

CHAPTER 11

Chester Andrew, Madera County Farm Bureau, California, speaks on the delegate floor at their state annual meeting.

(Photo by Vicky Boyd, California Farm Bureau.)

Value of Nominating Committees

Community and county Farm Bureaus usually select their candidates for office by means of nominating committees. This method of selection has decided advantages over the hit-or-miss method of proposing nominees on the spur of the moment from the floor. The advantages of a nominating committee are:

1. It can make a thorough study of the leadership needs of the organization for the coming year.
2. It can make a considered choice of nominees and can investigate the qualifications of each one.
3. It can interview prospective nominees, talk over the problems of the office with them, and give them time to think the matter over. It also avoids the confusing and damaging reaction which occurs when one member after another refuses to accept a nomination hastily thrust upon him from the floor. The morale of the whole group is lowered by such an incident.

   Actually, a member often declines a nomination because he is taken by surprise. He would probably accept if approached by a nominating committee which had considered all possibilities, was prepared to talk the matter over with him, and to allow him time to make a decision.
4. It can apportion candidates among different groups and localities.
5. It can choose nominees who will work easily and harmoniously with each other. It is good practice, after deciding upon a president or chairman, to consult him as to his preference for secretary. Experienced members who are nominees for offices or committee chairmanships can work effectively without being greatly influenced by their personal likes or dislikes. However, they can function even more effectively with some members than with others.

Selection of Nominating Committee

One of the fundamental principles of parliamentary procedure is that, where power is delegated by the members to an officer or committee, the selection of those who are to exercise that power should be made by democratic processes. If a president appoints the nominating committee, or serves on the committee himself, this principle is violated. A nominating committee should be a representative committee and the president should not be a member of it or meet with it.

In community Farm Bureaus, usually some of the members of the nominating committee are selected by the board and some are elected by the membership.

In counties, the committee is usually chosen from recommendations made by the county board of directors. The committee is composed in part of members who are not on a board.

The nominating committee in a community Farm Bureau reports to the membership, and in the county to the membership at the annual meeting.

Duties of Nominating Committee

The committee usually meets, with only its members present, to make a thorough study of the needs of the organization and of the qualifications of prospective nominees. Its meetings should be held early enough to allow the committee ample time to finish its work. It should consider the leadership needs of the group; determine skills or experiences needed by potential nominees; weigh the abilities of members for specific jobs; interview those who have been tentatively selected; agree to, vote on and submit a report.

The report should be submitted in time to send it out to members before the election date. Nominations which are to be voted upon by a convention should be sent out early enough to allow delegates to talk with the groups which they represent and to learn their opinions and preferences. Also, at a convention, delegates should have the opportunity to see and hear the nominees, if they wish, and to question them on their beliefs.

If a nominating committee takes its duties so lightly that it gets together ten minutes before the meeting at which it is to report and hastily assembles a list of nominees, it is doing immeasurable harm. The value of the nominating committee lies in the study which it makes and in the wisdom of the decisions which it reaches.

The committee often selects only one nominee for each office. There is nothing particularly "democratic" or advantageous about putting up additional nominees only to defeat them. It is sensible for a nominating committee to pick the one best candidate for each office, and not to pad the report with additional names. It is always possible for nominations from the floor to be added to the list proposed by the committee. Members should nominate from the floor if the nominating committee has not made a wise selection of candidates.

Report of the Nominating Committee

Although it is a common belief that no member who is on the nominating committee can be nominated for office by the committee, this is not true. The committee may nominate one or more of its members for office. If this were not possible, appointment to the nominating committee would bar a member from becoming a candidate, and excellent material might thus be lost. Also members favoring a nominee might eliminate possible rivals by getting them on the nominating committee.

However, it is unethical for a nominating committee to present the names of a considerable number of its members as candidates. For this reason it is good practice not to include on the nominating committee those members who are likely to be chosen as candidates.

The final report of the nominating committee must be voted upon at a meeting of the committee, and signed by the chairman. The chairman reads the report, when it is called for by the presiding officer, and hands it to the secretary. No action is taken on the report.

When the report of the nominating committee has been read, the presiding officer reads the title of each office separately and asks if there are additional nominees for that position. If he is wise, he bends over backwards to make it clear that no one is "railroading through a slate." By inviting additional nominations, and by waiting long enough after asking for other nominations, he gives members ample opportunity to propose names if they wish.

Closing nominations

When all members have had an opportunity to propose nominees, the presiding officer may declare the nominations closed, or a motion to close nominations may be made.

Members may vote for any eligible member for an office, regardless of whether or not he has been nominated either by the committee or from the floor, by writing in his name on the ballot. If a member receives the numbers of votes necessary, she is elected.

Electing officers

Usually, the bylaws specify that the officers are elected by ballot vote. If so, this provision must be complied with, even though the nominating committee's list presents only one candidate for each office and no nominations have been made from the floor. A motion to authorize the secretary to cast a ballot for the candidates violates the provision of the bylaws requiring a ballot vote.

The essence of ballot voting is the protection of the right of each member to keep his opinions and vote secret. No member can oppose or vote against a motion for the casting of a ballot by the secretary without being forced to reveal her opposition to one or more of the candidates. Even though voting by ballot consumes time, this requirement should be observed.

Ballots are prepared ahead of time. All offices may be voted for on the same ballot. Ballots may consist of uniform slips of paper, on which members write the name of the office and the name of their chosen candidate, or of a printed list of nominees.

At least three tellers are appointed by the presiding officer to pass out and collect the ballots. Any member has the privilege of watching the ballots counted. Votes which are cast for ineligible persons and ballots which are illegible or blank, are not counted as legal votes.

Unless there is a different provision in the bylaws, a majority of the legal votes cast is sufficient to elect to office. A motion to make a vote unanimous has no legal effect and would itself require a unanimous vote.

An election becomes effective immediately unless some other time is specified in the bylaws. Installation of officers is a ceremony which varies according to the preferences of the group. Officers are installed immediately after election, unless the bylaws provide differently.

Vacancies may be filled by election at any regular or special meeting, provided proper notice has been given, unless the bylaws specify a different method.

Term of Office and Reelection

Farm Bureau believes that elections should be held annually. This annual selection of leaders is important because it enables members to review the organization's needs and progress, to relieve those leaders who cannot continue or who have not measured up to their duties, and to select those whose leadership is needed by their group.

In general, Farm Bureau does not limit the number of times that an officer may be elected. In many organizations there is a fallacious belief that offices exist for the purpose of giving every member a chance to hold an office. Many groups believe that every member, at some time or another, should be president. Consequently, they limit the term of office or the number of times that an officer may be reelected, in order to "give everybody a turn." This is done with the mistaken idea that they are being "democratic" and preventing over-ambitious persons from controlling the leadership of the group. Actually, offices exist to promote the welfare of the organization and not to give members a chance to hold office.

Placing a limitation on the number of times that an officer may be reelected does not primarily limit the officer. Rather it limits the members or delegates and prevents them from voting for a member whom they may wish to continue in office.

CHAPTER 12: Committees

County Farm Bureau presidents work on county resolutions during a subcommittee meeting of the Arkansas Farm Bureau resolutions committee. State Board member Kenneth May headed this subcommittee.

(Photo courtesy Arkansas Farm Bureau.)

Committees are Important

Farm Bureau committees are working units of the organization. They actually do the bulk of the work that Farm Bureau accomplishes. It is important that every member understand how to work well on committees because committee service is deeply rewarding both to the individual and to Farm Bureau.

Committee work also trains Farm Bureau workers and leaders. Members who hope to become leaders, young members who are becoming active in Farm Bureau, and all who want to help with Farm Bureau's work find committee service a valuable experience and preparation for higher office, as well as an opportunity to do work for which they may be especially suited.

Some organizations lower the morale and potential of their committees by assuming that members dislike committee work, and committees waste time, that they are a necessary nuisance, or that members should be urged and coaxed to serve on them. Effective committee work is impossible when these mistaken beliefs prevail. Appointment to a committee is an honor and a privilege.

Farm Bureau knows that committees are the most important working force of the organization, that they can be efficient and effective, and that selection as a committee member is a genuine honor because it proves that your fellow members consider you both capable and dependable. Consequently, members seek service on Farm Bureau committees and often compete for places on important committees.

It is through your work on committees that you give your greatest service to Farm Bureau. It is through committee work that you get real understanding of your organization.

Advantages of Committees

Some groups make the mistake of trying to work out problems in meetings, which could better be handled by a committee – for a committee has advantages over a general group meeting. Here are some of them:

1. A small number of people can consider, plan and carry out a piece of work more efficiently than a large group. Let's keep committees small. Trying to get "everyone on a committee" is a mistake.

2. A committee can work quietly and with a calmness that is impossible in a meeting.

3. A committee can discuss more freely because procedure is more informal. Although discussion must be limited to one main subject and only one member may talk at a time, committees have little need for applying other rules strictly. Members may speak as often as they wish and may explain additional related ideas. For example, if you are discussing county zoning in a committee, you might also consider its effect on tax assessments.

4. A committee is usually made up largely of members who are personally interested in the problem being considered; in a larger meeting many of the members may be uninterested and unfamiliar with the problem.

5. In a committee, embarrassing, delicate, or controversial subjects can be talked over without publicity.

6. A committee may hold hearings and bring in outside authorities as consultants. For example, the county Policy Development Committee might arrange for a meeting with the president of the county Medical Association in connection with a proposed policy on health.

Standing Committees

A standing committee is one which stands ready to handle routine matters and problems connected with a specific part of the organization's regular work. Its members usually serve for the same term as do the officers of the organization. Standing committees are provided for in the bylaws. They are often appointed by the presiding officer, by the board of directors, or by both in consultation.

In some groups important standing committees are given even longer terms of office, or their membership is staggered from year to year, so that there will always be someone on the committee who is familiar with its long-term plans.

Usually only a few standing committees are needed. Most matters are best handled by special committees. Standing committees of Farm Bureau often include Membership, Policy Development, County and State Legislative Affairs, National Legislative Affairs, Women's Committee, Young Farmers' and Ranchers' Committee, Member Participation, Public Information, Commodity Advisory, and Promotion and Education.

Allen County Farm Bureau, Kansas, recruits members at a local gathering.

(Photo courtesy Kansas Farm Bureau.)

Special Committees

A special committee is selected to do a particular job and as soon as that job is finished, the committee is dissolved. Special committees may be appointed by the presiding officer, elected by the members, chosen by the board, or selected by a combination of these methods.

Choosing Committee Members

To choose the members of a committee intelligently, you need to know what type of work the committee is expected to do. Some committees should be representative committees – others should be composed solely of members who favor the project to be carried out.

A county Farm Bureau Board of Directors must be representative; that is, it should include a member from each section of the county, and frequently a member from each of the important groups so that it is representative of the major parts and interests of the county.

A committee appointed to recommend the location of a new bridge, for example, should also be a representative committee – that is, it should represent in its membership different opinions and different areas. If all of the committee members live on the south side of the county, those in other areas are not likely to agree with the report, and the committee's time will have been wasted. Whenever a committee is to investigate, recommend, study, or hold hearings, it should be a representative committee and its membership should reflect all important groups and viewpoints.

A committee to carry out a particular piece of work should be composed entirely of those who favor the work to be done. If some member is opposed to picnics, you wouldn't put him on the picnic committee. If you have a member who has tremendous enthusiasm for the committee project, put him on the committee and you will probably develop a new leader.

Don't appoint members to committees "to get them interested." Some members dislike committee work; they should not be urged to serve. Never appoint a committee until there is a job for it to do. Inactive committees lose interest.

If a member cannot be active on a committee, let him resign and give another member a chance to take his place. Any committee member who fails to do his work may be removed in the same way he was chosen. Too often this is not done. It is not fair to the other members to retain a member on the committee who does not carry his share of the load.

It is not always a good practice to ask the chairman of a committee to suggest names of members who he would like to have work with him. A chairman selecting his own committee would have the opportunity to select people who are biased towards their point of view. If the committee has a controversial assignment or issue to deal with, it is important that a good representation of the membership is included on the committee.

THE COMMITTEE CHAIRMAN

The president usually names the chairman of a committee at the time he appoints the committee. There is no reason why a member should be named chairman of a committee merely because she proposes an idea or moves that the committee be chosen. Proposing an idea does not prove that the proposer has the qualifications of a good chairman, or even of a good committee member. Neither should it bar her from being chairman if she does have the qualifications.

A committee chairman should be one who likes to work closely with people. She needs to have the ability to plan work and to delegate authority to other members. She should be someone who believes in the ability to plan work and to delegate authority to other members. She should be someone who believes in the ability of a group to make decisions, and not one who wants to decide and do everything herself.

The chairman is responsible for seeing that the committee works progressively toward its main purpose. The wise chairman recognizes, however, that individual members sometimes have their own legitimate personal interests in serving on a committee, in addition to their wish to help Farm Bureau; he plans the work so that they may satisfy some of these interests. For example, if Arnold Hicks wants to become better acquainted with his fellow apricot growers, the competent committee chairman gives him a job which will allow him to call on them instead of asking him to serve as secretary of the committee.

A good committee chairman seeks to have his committee work effectively as a unit. To do this he should be able to explain the work to be done, outline plans for accomplishing it, clear up misunderstandings, resolve differences, and use the abilities of each member to best advantage.

The chairman calls meetings of his committee unless regular dates have been fixed otherwise. If the chairman is unable, or for any reason fails, to call a meeting, any two members of the committee may do so. The president may request the chairman to call a meeting. No committee meeting is legal, however, unless all members are notified of it.

Tools of Committees

Every committee needs its tools for the first meeting, otherwise time will be wasted. What are the tools of a committee? The following are the minimum with which it should start work:

1. A list of the committee members, with addresses and telephone numbers – a copy for each member.

2. A statement, in writing, of the motion, problem, or piece of work referred to the committee. This statement should be definite because individual members often have different ideas about what the committee's job really is.

3. Instructions to the committee, from the organization or the president, explaining the committee assignment. The instructions include what the committee is expected to do, and how and when it is expected to do it. For example, is it to study a problem, hold hearings, investigate, make recommendations, or arrange a program? When should it begin work? When should progress reports be made? When is a final report due? Additional instructions may also be given a committee by vote of the members or by the presiding officer during a meeting.

4. A written statement of the powers of the committee. It is a common mistake to suppose that the mere appointment of a committee gives it definite powers. No committee has any authority except that specifically given to it by the organization or by the bylaws. Even a board of directors has no powers except those listed in the bylaws or voted by a meeting or convention. Has the committee the right to spend money? If so, how much? Can it conclude arrangements with other organizations? All such questions are answered in the statement of the committee powers.

5. Any material such as documents, letters, or records belonging to the organization which may help the committee in carrying out its work. If the committee is to arrange an annual banquet, the reports of the committees in charge of several recent banquets will be helpful.

6. Copies of any rules, plans, policies, or decision of Farm Bureau affecting the subject referred to the committee.

Most of this information may be found in the bylaws, policies, rules, minutes, records, or precedents of Farm Bureau. If there is a staff member assisting the group, he may provide these tools. If not, the president and the secretary of the organization have the responsibility to collect them. If they neglect this duty, the chairman of the committee gathers his own tools.

Ex Officio Members of Committees

Some Farm Bureau groups provide in their bylaws that the president or some other officer is an ex officio member of certain committees. In other words, he is a committee member by reason of holding the office. This enables officers to keep in touch with committee work.

An ex officio member of a committee has exactly the same rights, duties and responsibilities as any other member, including the right to vote, unless the bylaws say differently.

Procedure in Committee Meetings

If the committee is a large or important one, a secretary is generally chosen by the committee members. He notifies members of meetings and keeps the minutes of the committee. These minutes should be more complete than those of the organization because they often form the basis for the committee's report. They belong exclusively to the committee, and no one can demand to see the committee minutes unless he is a member of the committee. If a committee is holding hearings, the viewpoints of members who appear should be recorded.

The chairman of a committee, unlike the chairman of a meeting, may propose motions, speak on them and vote just as any other member.

In large committees, or boards, such as some county Farm Bureau boards of directors, it is necessary to follow procedure rather carefully, but small committees can be quite informal. No seconds to motions are required in any committee.

A majority of the committee members must be present in order to transact business legally, unless the bylaws fix a different quorum.

Any committee has the right to appoint subcommittees which report only to it.

The final report of a committee must be agreed upon by a vote of the members at a meeting of the committee. Voting by telephone or letter is not legal unless the bylaws provide for such methods.

Committee Reports

(Photo courtesy Arkansas Farm Bureau.)

Ashley County Farm Bureau President Bob Davis of Hamburg, Arkansas, listens during the discussion of Farm Bureau policy in the rice workshop of a district Farm Bureau policy development kickoff meeting.

Preparing a Committee Report

Good committee work deserves a good report. Unless a report is well prepared, members of the organization often fail to appreciate or understand the work that the committee has accomplished.

The length of a report varies greatly. A committee to make an important study may submit a fairly lengthy report, because it should include not only the results of the study but also the methods used and the reasons for the conclusions.

A committee to do a specific piece of work may submit a report of only a few lines. For example, a committee to revise a resolution may simply submit the revised resolution.

A committee report is usually worked out cooperatively by the members under the guidance of the chair. Since a report should represent the collective judgement of the majority of the committee, the members should have an opportunity to discuss the report with each other. It is necessary, therefore, that the report be agreed upon at the meeting of the committee and that a motion approving it as the report of the committee be voted upon.

When the report has been voted upon by the committee, it is signed by the chair.

(Photo courtesy Montana Farm Bureau.)

Jim Stenbeisser, Richland County, and Leroy Gable, Yellowstone County, listen to a report during a meeting of the Montana Farm Bureau Sugar Advisory Committee.

Contents of a Committee Report

A committee report should contain the following:

1. A statement of the motion, subject, or work referred to the committee, and the purpose for which it was referred; also any important instructions given the committee.
2. A brief summary of the plan or methods followed.
3. A summary of the work accomplished or the information gathered.
4. The conclusions of the committee.

If a committee wishes to propose recommendations, these should not be included in the main report but should be written on a separate sheet of paper. Each recommendation must be voted on individually.

Presenting a Committee Report

The chair, or any member selected by the committee, may present the report.

All reports, except routine progress reports, should be in writing. As soon as the report is read, or in the case of a long report as soon as it has been summarized, it is handed to the secretary.

If a substantial number of the committee members disagree with the report or any of the recommendations agreed upon by the majority, they may submit another report. Such a minority report is read immediately following the reading of the majority report. A minority report cannot be considered or discussed by the meeting, however, unless the members vote to substitute it for the majority report.

Acting on a Committee Report

When a committee report has been read, the meeting may act on it in any of the following ways:

1. The committee report may be **filed**. This means that the organization is including the report in its records. When a report is filed, the organization is not bound by it. It is not necessary for the assembly to vote on filing routine or progress reports. The presiding officer may say, "The report will be filed," and continue with the business.

    If the group wishes to thank the committee, this is done by a motion.

2. Decision on the report may be **postponed** until a more convenient time, by a motion to postpone to a definite time.

3. A report may be **returned to the committee** for more information or work.

4. A report may be **adopted or accepted**. If a committee report is adopted or accepted, the organization binds itself to any opinions, conclusions, or recommendations which are included in the body of the report. For this reason organizations should be cautious in accepting or adopting reports. Usually it is better to file them.

5. A committee report may be **rejected**. The organization may accept parts of the report, but reject others. The group cannot change the report, however, because it cannot make the report say what the committee does not wish it to say. Therefore no committee report can be amended from the floor. Recommendations presented separately from the report may be amended just as any main motion.

6. A committee report may be **referred** to an officer, board, or another committee for study. Financial reports are usually referred automatically to the auditing committee.

13

The final decision on the work, investigation, or conclusions of a committee belongs to the whole membership. The group, however, usually agrees with the committee and frequently adopts its recommendations as submitted. This fact emphasizes the power and value of committees.

(Photo courtesy Illinois Farm Bureau.)

Knox County, Illinois, Farm Bureau members take pride in their office and the flag that flies above it.

County Program Planning — The Process

CHAPTER 14

(Photo courtesy Iowa Farm Bureau.)

Iowa Farm Bureau members consider the programs they offer at a county Farm Bureau meeting.

14

COUNTY PROGRAM PLANNING – THE PROCESS

How It Started

It was nearing midnight on a warm August evening as the Board of Directors of Jefferson County Farm Bureau struggled with the few last problems. The directors had plowed determinedly through a lot of business. They had set dates for the county Farm Bureau annual meeting, selected the members for the Resolutions Committee, and developed plans for stimulating attendance at the state Farm Bureau convention in November.

They had spent hours working on policy recommendations received from all of the community Farm Bureaus. They realized that they had skipped, or only touched lightly upon, many items of the agenda. Most of the members would have to be up at dawn.

"Why do we always get into this kind of jam? Here we are almost at the close of the year, and how in the world are we going to get everything done?" queried Fred Lucas of Plainville Farm Bureau in a weary and discouraged tone.

There was silence until Kenneth Hardy from the Dallas Community Farm Bureau spoke up. "We simply have too much going on or else our program is getting too big. After all, we ought to recognize that even as members of this Board, we have to do a little farming on the side. We can't devote all our time to Farm Bureau."

John Stauffer from the Rocky Creek Farm Bureau leaned forward, elbows on the table, as he said earnestly, "I recall a few years ago when we had only two organized community Farm Bureaus in this county. Two years ago we started our community group. Then we began commodity activities. Our Young Farmers program has expanded tremendously, and we're mushrooming in all directions."

"The facts are," John continued, "when we look back just five years and compare the Jefferson County Farm Bureau of that day with what it is today, we all feel proud."

"Remember the state-wide training school several of us attended?" queried Homer Wilson, eagerly. "The course in program planning I was in stressed the importance – almost the necessity – of each county Farm Bureau developing a yearly program of work for the whole county."

At midnight, discussion on the program planning was interrupted by Abner Coxwell. "Mr. President, after listening to this discussion I have decided that, even though we do have a good program under way in this county and all of us are proud of it, we need something **better**. I move that the President appoint a program planning committee for next year, to be composed of the chair of each of the five community Farm Bureaus, Women's Committee, Young Farmers Committee, and of each county activity committee – and the county officers."

There were about four seconds to the motion and Herb Nolan called for discussion.

"What about our state field representative and the office manager?" questioned Wayne Carroll.

"It seems to me," said Abner, "that they should not be named on this committee, but it should be a part of their responsibilities to work with this committee just as they do with all other committees in the county."

The motion passed unanimously.

"Now," said President Nolan, "with your approval, I'd like to name as chairman a member of our Board who's helped on virtually every new project we've started during the past few years; he's also a former voting delegate from this county. I name John Stauffer as Chairman of our new Program Planning Committee."

There was a round of applause and President Nolan smiled as he continued. "Well, I guess you folks are in favor of that.

John and I'll get together at the office tomorrow morning, outline the assignments for the members of this committee, and arrange to get in touch with them about further plans."

"Mr. President," said Martin Gould, "this board meeting was scheduled for September 12. It's now September 13, and we should either stop the clock or adjourn. I move we adjourn." "Second the motion!" yelled everybody as the directors, exhausted but pleased with their new plan, gathered up their papers.

The President and the Chairman met at the office the next morning. They selected October 8 for the first meeting of the Program Planning Committee. Bernie Bishop, state Field Representative, and John Stauffer, Chairman, telephoned or visited each member of the Committee. Plans were readied, reminder notices were sent out, reference material was collected, and last-minute calls were made. The Chairman urged each committee member to give thought to the program needs of the county, and to come to the first meeting with definite ideas and suggestions.

How it Worked

Let's listen in on the first meeting of the Program Planning Committee of the Jefferson County Farm Bureau.

Talking animatedly, as they sit comfortably around the long oak table in the board room waiting for the meeting to begin, are the twenty-one members who make up the Program Planning Committee. The arrangement of scarlet and gold leaves, as well as the logs crackling cheerily in the fireplace, are just right for this crisp evening of October.

Present, in addition to John Stauffer, Chairman of the Program Planning Committee, are the following:

COUNTY OFFICERS
    President of Jefferson County Farm Bureau – Herb Nolan
    Vice-President and Resolutions Committee Chairman – Michael Ford

COUNTY PROGRAM PLANNING – THE PROCESS

Delegate to state Farm Bureau – James Mason

GROUP CHAIRMEN
- Plainville Community Farm Bureau – Fred Lucas
- Shady Glen Community Farm Bureau – Grover Turnbow
- Pinole Community Farm Bureau – Henry Schmidt
- Dallas Community Farm Bureau – Ken Harding
- Rocky Creek Community Farm Bureau – Carl Peterson
- Young Farmers Committee – Mario Guidi
- Women's Committee – Mrs. Doris Reed

COUNTY ACTIVITY CHAIRMEN
- Commodity Advisory Committee – Wayne Carroll
- Membership Committee – Dan Cleeves
- Policy Development Committee – Arthur King
- National Legislative Affairs Committee – Matthew Kitchin
- County and State Legislative Affairs Committee – Kenneth Holton
- Public Information Committee – Martin Gould
- Member Participation Committee – Jack Kessler
- Service-to-Members Committee – Abner Coxwell
- Budget Committee – John McTavish

Two Farm Bureau staff people are also present:
- State Field Representative – Bernie Bishop
- Office Manager – Rosemary Martin

Promptly at seven-thirty, Chairman John Stauffer rapped for attention.

"Every member of our Committee is here tonight, with one exception," he announced jovially. "Art King telephoned me last night that he couldn't be here until eight.

"By the way, I'm not sure that all of you know Jack Kessler, our new Member Participation Chairman. He's taken over the old Havens farm recently. We're asking him to help us do some of the things he worked out so well in Bates County Farm Bureau. Jack, we welcome you as a member of this Committee.

"Suppose we tackle our job this way. First, I'll outline our aim. You know, we've had a lot of difficulty in Jefferson County every year with conflicting meetings, with different Farm Bureau groups working on different ideas, and with some groups failing to cover the whole program of work. Even the Board of Directors is swamped due to these problems. Many important matters have gone by default because we haven't had plans set up for them. We've had confusion galore. We've certainly learned the hard way that we need a **planned program for the whole county for the whole year**.

Florida Farm Bureau Young Farmers and Ranchers recruiting new members at a rally.

(Photo courtesy Florida Farm Bureau.)

"Your county Board of Directors believes that this Committee can solve a lot of problems by joint planning in advance. I'm not mentioning any names," he continued with a twinkle, "but I was at one community meeting last year where the chairman asked, 'Has anybody got a good idea for the next meeting?' Let's start tonight to prove that county-wide, year-long program planning pays. Do you go along with this idea?"

"One hundred percent! It would sure do away with confusion and this last-minute stuff. It would help in getting new members, if we could show 'em just what they can look forward to for the year," spoke up Fred Lucas, Chairman of Plainville Farm Bureau, approvingly.

"It'll certainly give us a boost in getting newspaper space and radio time," commented Martin Gould.

"It'll make everybody's work easier," summed up Matthew Kitchin.

COUNTY PROGRAM PLANNING – THE PROCESS

"Right," said the Chairman. "We all seem to agree on the need for doing this job. Come on in, Art. We've kept a seat for you. After talking with all of you individually, I have a plan to suggest. This is it.

"This committee here tonight is composed of (1) those who represent a Farm Bureau group and (2) those who represent a Farm Bureau activity. The groups represented are:

1. The county Farm Bureau
   The five community Farm Bureaus
2. The Women's Committee
   The Young Farmers Committee
   The Commodity Advisory Committee

"First, this Committee here tonight will plan the county-wide meetings. Then each of the Farm Bureau chairmen representing a group will act as chair of a subcommittee to which other members, not on this Committee, will be added. Each subcommittee will fix the dates and plan the program for the year for all of the groups it represents. For example, Mrs. Adkinson, who represents the five community groups, will be Chair of our Subcommittee on Community Groups. She will get her Committee together, and they'll plan the program for the year for all five groups.

"When the plans of your subcommittee are complete, each of you subcommittee chairmen will report back to this Committee. These reports will be studied, approved and combined to form the report of this Program Planning Committee. That's the special job for you group chairmen.

"The rest of you are each chairmen of an important Farm Bureau **activity**:
Budget
Commodity Advisory
County and State Legislative Affairs
Member Participation
Membership
National Legislative Affairs

Policy Development
Public Information
Service to Members

"Each of the activity chairmen already has a committee. You and your committee members will act as advisers and assistants to the subcommittees.

"For example, take Sharon Holton, Chair of our Committee on County and State Legislative Affairs. Any subcommittee chairman may ask Sharon to help his or her committee plan a program on legislative affairs, suggest a speaker, or furnish material on that subject.

"The activity chairs also are responsible for seeing that your particular activity is represented on a program of each Farm Bureau group which is interested. We would like to spotlight and stimulate each of the major activities of Farm Bureau by having it on the program of all groups once during the year.

"It's also the duty of each activity chair to keep in close touch with group chairs through the year. Often emergencies arise in carrying out some policy. Then the activity chair concerned with that policy can help by speaking at meetings.

"In other words, the group chairs head the program planning for the year for your groups. The activity chairs lend a helping hand, whenever you can, to the group chairs and their subcommittees.

"If we succeed in this plan, every program and every meeting in Jefferson County will mesh. Together they'll make one complete program for the county for the year."

"Surely you don't mean that we'll actually sit here and help plan programs for the meetings of the Women's Committee, do you?" queried Ken Harding in mock apprehension.

"Certainly not, Ken," reassured John Stauffer, grinning. "You must have been snoozing for the last ten minutes. You won't be asked to plan a single one of the many important activities of our Women's Committee. They do an excellent job of that themselves.

"The Board of Directors has made up a list of the most important issues and projects which Farm Bureau should work on this year. Keeping these in mind, this committee will plan the programs for the county-wide meetings, and set the dates for them. Then each subcommittee will plan its own programs with the help and advice of the activity chair.

"For example, after we have talked over the important ideas and dates for this year's county-wide meetings, Mrs. Reed, Chair of the Women's Committee, will get her subcommittee together, and they will plan their own program for the year. As Chair, she will report back to this Committee at our next meeting. Likewise Carl Peterson, Chairman of Rocky Creek Farm Bureau, and his committee will work out their programs for the year along the general lines we agree upon, and report to this Committee. What do you think of this plan?"

"I'm for it! This over-all program planning for the whole year means a lot of work – but Jefferson County will get a lot of benefit from it," said Wayne Carroll.

"Just one word. I begin to think it might save us a good bit of money, too," said Budget Chairman John McTavish, cautiously. "If this Committee does the planning all at once, Farm Bureau will save in the long run. There won't be program committees wasting stamps all year, and there'll be time to write speakers instead of calling them long distance and running up bills."

"What appeals to me," enthused Grover Turnbow, Chairman of the Shady Glen Farm Bureau, "is that all our groups will be free to carry out their own particular ideas in their own way, yet we'll all be pulling together under the same general plan. No more conflict of dates – no more neglect of important issues – no more confusion."

"Bernie, as state Field Representative, what's your feeling about this?" queried Michael Ford.

"Mr. Chairman," replied Bernie Bishop, "I believe the work of this Committee can increase Farm Bureau's achievement tremendously. To me, it's a big step forward."

"Since we all understand our aim and plan of work, now let's consider our county-wide meetings," said Chairman Stauffer. "The county Farm Bureau is the smallest **uniform** unit of Farm Bureau administration all over this country. It is the strongest local group of Farm Bureau, and is close to every member.

"Concerted action should begin through the county Farm Bureau. All local groups get information and understanding of important issues from the county. The activities of each of our community groups, our discussion groups, our county committees – all center in the county Farm Bureau. It is the nucleus around which all community Farm Bureau activities revolve. It is the center of gravitation. **County Farm Bureaus, therefore, must be strengthened and kept strong.**

"Realizing this, your Board of Directors of Jefferson County recommends that more county-wide meetings be held.

"I suggest," said Vice-President Michael Ford, "That this year we try adding two more county-wide meetings – maybe the first one around February, when it won't interfere too much with planting in this county, and then another one in September, keeping away from harvest dates as much as possible.

"The February meeting could cover Farm Bureau policies with regard to state and national issues. This would also be a good meeting for launching any special projects. The meeting in September has many possibilities and one of our committee members has already discussed an idea with me which he'll give you later in this meeting.

"Our annual meeting would still come in late October as always. Through these three big meetings and other county-wide events, we'd really get to know each other and build a lot of strength and unity. I'm all for trying out the recommendations of the Board."

"Thank you, Michael," responded Chairman Stauffer. "The next…"

"Just one word," broke in John McTavish, agitatedly. "This all sounds good. But you are forgetting that more meetings are going to cost a lot more money – light, heat, notices, food – to say nothing of 'miscellaneous.'"

"Certainly," agreed Carl Peterson, "but won't the money be well spent? After all, the best way for Farm Bureau to get more members and more money is by having more good meetings."

"Seems to me we're doing all right now. Why try something new?" sputtered McTavish.

After considerable discussion, a motion approving the two additional county-wide meetings carried unanimously.

"Now," suggested Chairman Stauffer, "I think it's time for a break. Bernie and Ken, how about some milk and cookies while we all take a three-minute stretch?"

When the meeting continued, Dan Cleeves began with "Mr. Chairman, I got the idea during that stretch. When all the subcommittee reports are in and we get the dates together for all the Farm Bureau meetings in Jefferson County for the year, why can't we have them printed on an attractive calendar for every Farm Bureau home?"

"Just one word," interjected John McTavish, struggling to his feet in alarm. "That's bound to cost considerable money. We can't…"

"Wait a minute, Mac," interrupted Abner Coxwell, "I think our service companies and our insurance services would be glad to assist in such a plan. It would be an excellent advertising medium for them."

"Sounds good," commented Chairman Stauffer. "Abner, will you please look into it and report at our next meeting?"

"I think it's time to call on our President. He's got some ideas for you – President Nolan."

"Your Board of Directors is delighted to know that we will really have an over-all yearly plan for Farm Bureau activities of Jefferson County," began the President. "Like McTavish here, I want to give you 'just one word' about planning programs – the word is **balance**.

"Our meeting programs need balance between county, state and national projects. We also want balance between our important commodity interests – serving the dairyman, helping the cotton growers and assisting the fruit growers. Suppose that each commodity group were to work strictly for its own welfare. A conflict of interests would throw Farm Bureau's work **out of balance**. The various agricultural commodities share many important interests. So at times, we know that it is wise to

Tom Byers (right) and Steve Maple (center) serve ice cream and strawberries at an Indiana Farm Bureau district meeting.

(Photo courtesy Indiana Farm Bureau.)

compromise our smaller differences and work together for our greater common interest. Thus everyone benefits. This is the kind of balance we must strive for all through Farm Bureau.

"Likewise we want balance in our meetings so that we interest the whole family – something for Dad, something for Mom and something for the young folks. We need balance between information and entertainment – or call it work and play.

COUNTY PROGRAM PLANNING – THE PROCESS

Some of you may accuse me of being overly serious. But I don't think we should have too many meetings where Farm Bureau people get together and just have fun. People are disappointed when they come to a Farm Bureau meeting if they don't leave knowing more than they did when they came. Our meetings should offer facts to take home, as well as fun and food.

"Finally, we need balance between the activities of Farm Bureau and training our members to carry out these activities. A balanced program must not only provide the work to be done, but also the training in how to get that work done effectively.

"Balance in all those things is just as important as the balanced ration for hogs or cattle, or a balanced diet for humans – we need it and it works. So my advice is – keep the word **balance** in mind as you plan."

The President smiled as the committee members nodded vigorous agreement.

"Thank you, Herb, for those words of wisdom," said Chairman Stauffer. "Here's the list of the most important issues and projects for this year that your Board of Directors made up."

"Can we get copies of it?" queried Mrs. Adkinson eagerly.

"I knew you'd ask that," said the Chairman with a smugly pleased nod, "and I'm ready for you. As soon as I finish reading the list, Rosemary will pass out copies to each of you.

"Here are the issues:

COUNTY TOPICS

1. Rezoning of rural areas.
2. Reassessment of farm properties.
3. Safety program.
4. Better schools and higher salaries for teachers.
5. Development of better farm-to-market roads.
6. Development of either the Harden or Green Hills Dam site.

STATE TOPICS

1. Development of highways and secondary roads.
2. Increase in school facilities and better teacher training.
3. Farm research through our Experiment Stations, Extension Service, private industry and on the farm.
4. Health Education.
5. Development of land and water resources.
6. Development of additional sources of farm labor.

NATIONAL TOPICS

1. Development of International Trade.
2. A balanced budget, with readjustment of certain taxes.
3. A farm program.
4. Federal versus state highway development.
5. Dairy products promotion (plus other product promotion if needed).

"Keeping in mind these issues and our continuing activities, let's pick the dates and agree on program ideas for each important county-wide meeting, starting with January. Once these county-wide meeting dates are set, no other Farm Bureau meeting should be planned to conflict with them."

"Our county Leadership Conference comes in January," spoke up Ken Harding. "Has that date been fixed?"

"It's all set for January 18 to 20," answered the President.

"We mustn't forget," prompted Dan Cleeves, "that on January 27 the Extension Service has its big meeting to discuss its agricultural projects for the year. Usually several hundred Farm Bureau people go to that, so we should set aside that day."

"Good," responded Chairman Stauffer. "Got both those dates down, Rosemary?"

"I've been wondering whether March 16 wouldn't be an excellent time for the first one of our new county-wide meetings. If you like the date, I have a surprise for you. At our state convention, I checked with our state administrator. He can talk to us on March 15 on **Progress on Legislative Issues**. You know what an outstanding speaker Wayne is. I asked him to keep the date and, if you approve, I'd like to call him and confirm it. This meeting would certainly delight all of our policy execution committees as well as our membership in general, and it would be timely."

"Mr. Chairman," said Arthur King, his voice ringing with enthusiasm, "I move we grab our state administrator for March 16. We're proud of you for getting him because that's one of his busiest times. There won't be many counties that will have this opportunity."

After an approving vote, Mrs. Doris Reed waved her hand for recognition.

"The Women's Committee has set May 1 for their Public Information Luncheon," she announced proudly.

"Okay," said the Chairman, "We'll check off May 1."

"And May 14, too," spoke up Mario Guidi. "We want that day for our Young Farmers County-wide Program."

"Right," responded the Chairman. "Now during the summer I don't suppose we'll want a county-wide meeting, but that extra meeting in the fall – Yes, Grover..."

"Mr. Chairman," said Grover Turnbow. "I move that September 25 be reserved for a county-wide Public Information dinner and that we call it the 'Farmer-Businessman's Banquet.'"

After a hearty approving vote, the Chairman continued. "October 2 is the Membership Kick-off Meeting, isn't it?"

"No, no," shouted Dan Cleeves excitedly. "It's October 7."

"Thanks for the correction," said the Chairman. "I must be getting a little absent-minded. On October 18 the Dairy Producers are holding their annual Dairy Review, and the Poultry Producers meet October 21 for their annual banquet.

"The Board of Directors has set October 26 for the county annual meeting. We chose that date because it gives us time enough to discuss and approve our recommendations for policies in preparation for the state convention. Our state convention runs from November 2 to November 6. Then Field Crops Day and Fruit Growers Day both come this year on November 15 – then Livestock Day on November 18. Have we forgotten any county-wide dates?"

"You've forgotten one important date – the convention of American Farm Bureau Federation on January 7 to 12," prompted Mario Guidi. "I know lots of our members who are going."

"Thank you, Mario." said the Chairman. "Now if no one has any other county-wide dates to add, I'll read the list as it stands now.

DATES RESERVED BY JEFFERSON COUNTY FARM BUREAU

January 7-12	Convention of American Farm Bureau Federation
January 18-20	Leadership Conference
January 27	Extension Service Conference
March 16	Legislative Dinner
May 1	Women's Committee County-wide Day
May 14	Young Farmers Committee County-wide Day
September 25	Farmer-Businessman's Banquet
October 7	Membership Kick-Off Meeting
October 18	Dairy Producers Dairy Review
October 21	Poultry Producers Annual Banquet
October 26	County Annual Meeting
November 2-6	State Convention
November 15	Field Crops Day-Fruit Growers Day
November 18	Livestock Day

"So far," continued Chairman Stauffer, "we've agreed on the dates and nature of our county-wide meetings, and the dates reserved for state and national meetings. Our next step is setting up the subcommittees for the group chairmen."

I've talked with each of you group chairmen and together we've agreed on appointments for the subcommittees. The group chairmen have obtained the consent to serve of each member whose name I read.

"I announce the appointment of the following:

SUBCOMMITTEES OF THE PROGRAM PLANNING COMMITTEE OF JEFFERSON COUNTY

COMMUNITY FARM BUREAU SUBCOMMITTEE

Chairman – Ken Harding, Dallas Farm Bureau

Members – The chairman of each of the other community Farm Bureaus. Each community chairman may select as many members of his community Farm Bureau as he wishes to help him plan the year's meetings for his group.

WOMEN'S SUBCOMMITTEE

Chairman – Mrs. Doris Reed, Chair Women's Committee

Members – The chair of the Women's Committee of each community Farm Bureau

YOUNG FARMERS SUBCOMMITTEE

Chairman – Mario Guidi, Chairman Young Farmers Committee

Members – The chairman of each standing committee of the Young Farmers

COMMODITY ADVISORY SUBCOMMITTEE

Chairman – Wayne Carroll, Chairman Commodity Advisory Committee

Members – The chairman of each commodity group

Any questions," asked Chairman Stauffer.

"Yes, Mr. Chairman," said Henry Schmidt. "I think some explanation of the plans of the Member Participation Committee and the Public Information Committee would help all of us subcommittee chairmen to cooperate better with them."

"Certainly. Jack, as Chairman of the Member Participation Committee, would you summarize the plans of your Committee?'"

"Pardon me, Mr. Chairman," interrupted President Nolan. "May I say something before Jack gives his summary? I'd like to sort of explain the ideas behind this committee's work."

"Certainly, Herb, go ahead."

"The Member Participation Committee is undertaking a vital Farm Bureau job. If Farm Bureau is to reach its potential, every member needs the confidence that comes from knowing the rules of working together in a group.

"If Farm Bureau members are familiar with the principles and rules of making motions, proposing policy recommendations, taking part in discussion, speaking, voting and serving on committees, they will be comfortable and confident in taking part in Farm Bureau meetings. They'll also represent Farm Bureau creditably at other meetings."

"You know, Herb," interrupted Fred Lucas, "I can testify that what you are saying is true. Before I had some training in taking part in meetings, I was scared to death to open my mouth. Every time the chairman called for discussion, I'd slide down a little lower on my spine and pray nobody would call on me. Now I really enjoy taking part."

"Thanks, Fred, for the testimonial," said the President. "Leadership training is excellent, but not all of us can attend the leadership conferences. Furthermore, those who do go need opportunity to practice what we learn at the conference. They tell us there that leaders alone can't make an organization strong. Organization strength depends on informed and participating members.

"Farm Bureau has a great opportunity to familiarize its members with the simple practical principles of cooperation. The member who understands the rules can play the game. He is an interested, active participant – eager to do his share.

The member who isn't confident is at a disadvantage. He loses interest and stays home.

"Investment of a small amount of time, effort and money in training members will pay high dividends. If Jefferson County can give its members a clear understanding of the fundamental ideas and techniques of how to work together effectively in Farm Bureau, it will create strength of which we've never dreamed. Then too, members will have real equality of opportunity when they all have the tools for participating.

"That's what the Member Participating Committee plans to give Farm Bureau members. Thank you."

"I'm mighty grateful to you, President Nolan, for that explanation," said Jack Kessler. "Briefly, here is how we hoped to promote member participation this year. We'll add other ideas later.

"We plan a brief member participation program at the beginning of each meeting. The five community Farm Bureaus of Jefferson County, the Women's Committee and the Young Farmers Committee all plan to spend not less than ten, and not more than twenty minutes, at the beginning of each meeting in explaining and practicing ways to participate effectively in Farm Bureau's work.

"Some of these groups will start their meeting fifteen minutes earlier; some plan to streamline their business sessions a bit to get time for training.

"I have the subject outline for the year here. You'll each get a copy. If necessary, staff members will assist by acting as leaders the first few times. Then the members will take over, using material supplied by Farm Bureau and by our book, **Your Farm Bureau**.

"First, there will be a five-minute talk on some idea for developing policies, or ways of participating in meetings, or some suggestion on speaking or presiding, or working on committees. Next, there'll be a brief practice or discussion session, with everybody who wants to getting into the act.

"The practice sessions should be fun – not dull. We suggest you get away from farm problems so that you can concentrate better on the way to do the activity, not on the subject. If you practice as a Farm Bureau group, you'll be talking issues or policies, or some Farm Bureau subject which you know, and you'll get so interested in the subject that you'll forget about the techniques and methods. If you are pretending to be some other group, you will concentrate on the **method** of doing the thing.

"Let your practice group imagine that it is a meeting of the Petunia Association, the Society for the Study of Interplanetary Travel, or the Club for Retired Racing Drivers. Try to act the way the members of whatever group you choose to be might act, and discuss ideas which would be of interest to them."

"Now," said Ken Alexander, passing around copies, "what I have just passed out is typical of what we hope to have as an opener for every Farm Bureau meeting."

It's easy for anyone to lead a member participation session because much of the material is in **Your Farm Bureau, Robert's Rules of Parliamentary Procedure**, or in the many pamphlets which Farm Bureau sends out. Members can suggest additional topics and request the Farm Bureau county or state staff to furnish material for them."

"Thank you, Ken," said Chairman Stauffer. "That's a mighty fine summary of what we can do in training our members. With a little effort and fun at each meeting, we can make Farm Bureau members outstanding in any group to which they belong. Every Farm Bureau group can use some of the ideas of this Committee."

"Now, Martin Gould, will give us a summary of the plans of your Committee on Public Information?"

"John, may I speak again?" asked President Nolan. "This is another committee that is newer than some, and I'd like to explain briefly its place in Farm Bureau activities.

"Actually, everything Farm Bureau does helps to inform the public on what Farm Bureau stands for. In the broadest sense, every member of Farm Bureau is a one-man committee on public information. In fact, Farm Bureau's public information program is made up of its policies, its leaders, its committee members, its activities, its projects, and most important of all, the everyday work of its individual members. The Public Information Committee is only one phase of our organization's public information program.

"This Committee plans specific methods and events for informing those who are not members of Farm Bureau on the policies and work of our organization."

"Mr. President, I greatly appreciate your introduction," said Martin Gould. "I'll read my summary."

SUMMARY OF PLANS OF COMMITTEE ON PUBLIC INFORMATION OF JEFFERSON COUNTY FARM BUREAU

"This committee plans to work through the press, radio and television and also through face-to-face explanations of Farm Bureau.

1. PANEL

    We plan to set up a speakers' panel consisting of two county directors and our Vice-President. This panel will be on call to talk on radio or TV or before such groups as the Bankers' Association, the Church Council, sportsmen's clubs, service clubs, civic associations and, in fact – any group wanting to know more about Farm Bureau.

    The panel will speak on such subjects as:

    A. Issues Facing Agriculture
    B. What Farmers Believe about National Issues
    C. Farm Bureau and the Way It Works

2. SPEAKERS' BUREAU

    Six speakers will be on call to represent Farm Bureau by talking to interested groups. This year's speakers will be our county president, the Chairman of the Young Farmers, the Chairman of the Commodity Advisory Committee and two of our county directors.

3. EVENTS

We plan two important events for informing the public about Farm Bureau. The first will be the Farmer-Businessman's Banquet on September 25, featuring a talk by our state Farm Bureau president. Our guests will be the mayors of our three towns, members of the county Board of Supervisors, the Director of Extension, the editors, the radio directors and the presidents of other men's and women's groups of Jefferson County.

Each Farm Bureau member may invite one businessman as his personal guest. Following our President's talk, several of our guests will discuss Organizational Cooperation in Jefferson County. The program will be broadcast on Station KTM.

The second event will be the Women's Committee Public Information Luncheon to be held on May 1, which is the Women's Committee County-wide Day. I'll let the Chairman of the Women's Committee tell you more about this.

4. LEADERSHIP IN OTHER ORGANIZATIONS

Farm Bureau is constantly called upon to furnish members to serve on committees and boards for county and community activities, along with representatives of other organizations. The training planned by the Member Participation Committee will certainly help our Public Information Committee. Trained Farm Bureau members are in constant demand to lead church groups, Chamber of Commerce committees, PTA's, political, business, fraternal and service groups. Training and developing these skilled members and leaders is an excellent contribution to the public Information activities of Farm Bureau."

"That's a splendid report," commented Chairman Stauffer. "Now it's getting late. Unless someone has a better date to suggest, we meet again November 10 – almost a month from tonight. At that meeting we will hear and discuss our report so that I can present it to the Board of Directors."

"Mr. Chairman," said Jack Kessler, "I suggest that Rosemary keep a large program calendar in the county office so that each group can check its dates with her and clear them before submitting its report to this committee."

"Good idea," commented Chairman Stauffer. "Here are copies of the so-called 'Instructions' for each subcommittee. Actually they are facts for your guidance. Dan, will you please pass them out? Let's read them together and see if there are any questions."

INSTRUCTIONS TO SUBCOMMITTEES

1. REPORT DUE: November 10
2. TYPE OF REPORT: A written report showing the date and detailed program for each meeting for the year of each Farm Bureau group assigned to your subcommittee.
3. PROCEDURE: Get your committee together promptly. Present the ideas which we have agreed upon at this meeting, and work out your meeting programs.
4. AIDS: Provide each of the members of your committee with the following:
    A. List of committee members and their telephone numbers
    B. List of the chairmen and members of each county Farm Bureau activity committee
    C. List of meeting dates already reserved by Jefferson County
    D. List of important issues – county, state and national

"Rosemary, will you give each subcommittee chairman enough copies of these lists for each member of his subcommittee? Is there anything anyone wants to add?"

14

"I think we've done a big piece of work tonight," commented Grover Turnbow proudly, as Mario Guidi and Doris Reed came in from the kitchen with trays of coffee, milk and cookies.

"And the subcommittees have a big piece of work ahead to be ready for the next meeting," added Henry Schmidt, helping himself liberally as the cookies were passed a second time. "I'll need another handful of these cookies to build up my strength for the job. But when it's all done – boy, what a wonderful feeling!"

County Program Planning — The Report

Chapter 15

Arkansas State Senator Morril Harriman (left) visits with (counter clockwise, left to right) Harlan Brammer, Crawford County Farm Bureau vice president; James Strang, Sabastian County, state Farm Bureau board member; John McClurkin, Crawford County Farm Bureau president; and Rose Arnold, Crawford County Farm Bureau Information Chair. The visit took place during the Arkansas Farm Bureau "Farmers Day at the Legislature," when county leaders visited their respective legislators and concluded with a reception and dinner for the leaders and legislators.

(Photo courtesy Arkansas Farm Bureau.)

15

Plan Your Work — Work Your Plan

"Perfect attendance again tonight? My telephone calls must have done some good," called out Chairman Stauffer proudly as the committee members settled themselves comfortably and the November 10 meeting began. "I see that some of you overworked chairmen of subcommittees are still polishing up your reports. We've got to cover a lot of ground tonight – so let's start."

"Mario Guidi, as Chairman of the Subcommittee on Programs for the Young Farmers, will you report first?"

"Mr. Chairman," began Mario Guidi promptly. "Our committee met three evenings. We tried to plan programs which will have the balance suggested by President Nolan and which will meet the needs of young people and will enable them to make their full contribution to Farm Bureau.

"Just as soon as our program is approved, we plan to reproduce copies of it with cartoon graphics and send it with a letter to every member. As you know, we meet on the second Thursday of each month. Here's our program."

JANUARY
Chairmen: Sam and Arlene Saunders
- 8:00 Member Participation: **Stating Motions Clearly**
  Leader, Paul Mason
- 8:20 Business meeting
- 8:45 Member-Panel Discussion: **How Today's Agricultural Issues Affect Our Future** – Sam Broderick, Gladys Bull, Henry F. Peterson, Fred Rinklander
- 9:45 Social Hour – Malted Milk

FEBRUARY
Chairmen: Elliot and Adele Mosher
- 8:00 Member Participation: **Informal Discussion**
  Leader, Hugh Davenport, Jr.
- 8:20 Business Meeting
- 8:35 **Applying Farm Bureau Policies to Present Issues**
  Herb Nolan, President, Jefferson County Farm Bureau

9:15   Cake and Coffee

MARCH
Chairmen: Henry and Hilary Sofio
6:30   BOX LUNCH AUCTION – Auctioneer, Martin Lyon
7:30   Member Participation: **Choosing a Committee Chair**
7:50   Business Meeting
8:15   **Our Private Enterprise System** – John Fallis, Farm Program Director KYZN
       Question Period
9:15   Folk Dancing and refreshments

APRIL
Chairmen: Allen Kendrick and Millie Tolan
8:00   Member Participation: Skit **"I'll Never Forget That Committee Meeting!"** – Directed by Finn MacCumail
8:25   Business Meeting
8:40   **The International Scene Today** – Prof. John Borden, Chairman, Dept. of Economics, A&M College
       Discussion Period
9:30   Weenie Roast

MAY (Second Saturday)
Chairmen: Peter and Ann Holden
9:00   BUS TOUR OF HISTORIC SPOTS – Meet at Jefferson County Farm Bureau Building
12:00 Picnic Lunch at Lake Almaden – Bring musical instruments, bathing suits and lunch

JUNE
Chairmen: Folger and Elaine Evans
6:15   CHICKEN DINNER
7:15   Member Participation: **Steps in Presenting Motions** – Bernie Bishop, State Field Representative
7:45   Business Meeting
8:00   **What It Takes to Get Started In Farming** – Kraut Huth, State Farm Management Specialist
9:30   Dancing and refreshments

JULY (Second Saturday)
Chairmen: Craig and Cricket Dinsmore
12:00 PICNIC AND BEACH PARTY: Eel River Bend; milk, coffee and soda provided
       Bring musical instruments

AUGUST
Chairmen: Ken and Alice Sigruts

> The Young Farmers Committee will be in charge of the **Get Acquainted Booth** and **Exhibit** of Jefferson County Farm Bureau at the State Fair Grounds, August 9-15
> Open 10:00 a.m. to 11:00 p.m Daily

SEPTEMBER
Chairmen: Shirley Farr and Trigg Scott

8:00 Participation: **Qualities of a Good Presiding Officer** – Leader, Cherie Persian

8:20 Business Meeting

8:40 Debate: **"Resolved: That a New Farm Safety Program is Needed"** – Vladimer Waleska, Ethel Bluman
Discussion

9:00 Dancing and refreshments

OCTOBER
Chairmen: Harry Scott and Rosemary Wayne

8:00 Member Participation: **What Is A Good Member?** – Leader, Allen Van Camp

8:20 Finals for Discussion Meet Contest

10:00 Halloween Frolic

NOVEMBER
Chairmen: Morris and Alice Henderson

6:30 **THANKSGIVING BANQUET** – Guests: Editors, radio commentators, presidents of Junior Chamber of Commerce, Future Farmers and similar groups
Each member may invite one individual guest

8:00 **Putting Jefferson County on the Map** – Speakers: Mario Guidi, Chairman Young Farmers Committee of Farm Bureau and Guest Presidents

DECEMBER
Chairmen: Marilyn Field and Roger Edwards

8:15 **CHRISTMAS PARTY**
Singing, stunts, games, gifts, dancing, eats!

"Thank you for a fine report," commented Jack Kessler. "Don't all of you wish you were eligible for the Young Farmers Program?"

"Are there any questions on Mario's report?" asked Chairman Stauffer. "If not, Doris Reed, will you give your report as Chair of the Women's Committee?"

"Mr. Chairman – the Women's Committee plans two important meetings for this year. The first is an Urban-Rural Conference and Luncheon to be held on October 9. The second is a Public Information Luncheon to be held on May 1.

"On the recommendation of the Board of Directors, the Women's Committee has decided to undertake a survey on Traffic Safety. We plan to make this survey in cooperation with the following agencies of Jefferson County: Department of Highways, Sheriff's Office, County Engineer, Department of Motor Vehicles, the Highway Patrol. We will also ask the cooperation of the various civic clubs and town councils.

"The Women's Committee will first hold a meeting of all the chairs of the community Farm Bureau women's committees and of the safety committees to explain this project. Next, each community and safety chair will assist with observations and special surveys of traffic in her community for two months. At the end of this period, these chairs will hold another meeting with the Women's Committee to report the facts they have gathered.

"The Women's Committee will then prepare a final report for the Board of Directors. It will also inform citizens on the report at a county-wide, Rural-Urban Women's Conference.

"Each member of the Women's Committee and each chair of the Women's Committee of the five community Farm Bureaus will be asked to suggest the name of a leader of a women's organization in her community to be invited to the conference. County presidents of the various women's organizations will also be invited. Invitations, signed by the Chair of the Women's Committee, will be sent out by the Farm Bureau office.

"The conference will start with a luncheon and each Farm Bureau woman will have an urban woman as her partner.

"After luncheon, there will be talks on **Our Findings on Traffic Hazards in Our County**, by the Chair of the Committee and by some of the members. Maps, marked to show situations in the county where there are traffic hazards, and charts illustrating the entire traffic-safety problem will be explained.

"This traffic survey of Jefferson County should throw light on the following:

1. Main highway traffic through towns – route of the highways, speed and warning signs.
2. Number of accidents – location, cause and fatalities.
3. Location of traffic signs and lights.
4. How our state and county rank in safety, by comparison with other states and counties.
5. Recommendations for improving traffic safety in the county.
6. How rural and urban people can work together to promote traffic safety.

"News releases describing the aims, progress, conclusions and recommendations of the committee's survey will be prepared for the papers throughout the county. Several women will participate in radio broadcasts.

"The chair of the women's committee of each of the five community Farm Bureaus will explain the survey and its results at local PTA meetings, civic groups and service clubs.

"We are convinced that this survey of traffic safety in Jefferson County will be a valuable contribution. We believe that the Urban-Rural Luncheon will make the survey known and will result in action on the committee's recommendations.

Congresswoman Olympia Snowe discusses issues with Young Farmers at Boothby's Century Elm Farm in Livermore, Maine. From left to right are Jon Olson, Jeff Bragg, Kyle Price, Senator Snowe, Clint Boothby and Rob Boothby.

(Photo courtesy Maine Farm Bureau.)

"The Women's Committee also plans a second event – a Public Information Luncheon on May 1. Our state chair of the Women's Committee will speak on **Our Mutual Responsibilities in Strengthening America**."

"We have also set up plans and committees for the following Farm Bureau projects:

1. Establish at least five more neighborhood groups in the county this year.
2. Carry out special assignments from the Membership Committee.
3. Work with the Commodity Advisory Committee on marketing and promotional programs.
4. Assist in gathering information for policy development meeting.
5. Participate in the five-point program of our state Women's Committee which includes:
   A. International understanding and trade.
   B. Citizenship.
   C. Rural Health.
   D. Market development.
   E. Consumer relations.

"The Women's Committee will strive increasingly, both as a group and as individuals, to help with all projects of Farm Bureau."

>Doris Reed, Chair
>Women's Committee of Jefferson County Farm Bureau

"Thank you, Doris Reed. We all appreciate the fine work and cooperative spirit shown by the Women's Committee. Are there any questions or comments on this report?

"Now we'll hear from the Commodity Chairman, Wayne Carroll."

COUNTY PROGRAM PLANNING – THE REPORT

"Mr. Chairman," began Wayne Carroll, "the Subcommittee on Commodities met twice and talked over plans. Each member of the subcommittee chose two other persons from his commodity group to work with him.

"Each of the larger commodity groups has planned a county-wide promotion day. You heard the dates read in the list of county-wide meetings.

"I am sure that all of you realize that many of the interests of each commodity group are so highly specialized that it is impossible for anyone except a member of the group to judge their suitability. Also, most of the subjects chosen for their meetings are so technical that this committee would not even understand some of them.

"Consequently, I asked my Commodity Committee to assume responsibility for approving program ideas and subjects for all the commodity groups. The committee has gone over each program in detail, and I assure you that the programs are all well planned.

"Each commodity group was careful to avoid any conflict of dates with any county-wide meeting, and we have cleared the dates for all of our meetings with the Farm Bureau office.

"Rather than report in detail on the technical subjects to be covered at most of the programs, I have decided to report the general program plan of each of the leading commodity advisory groups.

1. The **Poultry Producers** will meet on the second Tuesday of each month. All of their programs are concerned with marketing and meat inspection.

2. The **Livestock Producers** are holding three product-promotion meetings. At present they are extremely concerned about marketing problems and these problems will be the subject of all three meetings.

3. The **Dairy Producers** are holding three meetings. All of their meetings will be devoted to the subjects of

increasing the consumption of dairy products and the more efficient production of these products.

4. The **Fruit Growers** plan four meetings. Each of their meetings centers around two subjects – grading and standards, and the increasing problem of pest control.

5. The **Field Crop Growers** have decided on three meetings. Their first program will deal with the problem of surplus production; the second, pest control; and the third with grades and standards.

"Those of you on this committee who have worked closely with the various commodity advisory committees, realize that there will be other commodity group meetings. There are some commodity problems which you cannot sit down and project ahead. These problems arise as emergencies, and need prompt, close attention.

"Flood conditions, and outbreak of disease, a new pest, an acute local market problem due to dislocation in the marketing pattern – these and many other emergencies would call for special meetings of particular commodity groups."

Wayne Carroll, Chairman

"Thank you, Wayne. I think everybody on this committee is pleased that your groups have completed their programs so efficiently. Frankly, I'm not at all sorry to pass the responsibility to your committee for approving such technical topics as **Eradication of the Alfalfa Aphid** and to skip the implications of the highly controversial topic of **Federal Meat Grading**.

"It is plain from your report that all of us will be eating twice as much food as we ate last year. Those commodity promotion meetings get results.

"Ken Harding, will you report for the community Farm Bureaus?"

"Mr. Chairman, each of the five community Farm Bureaus has a well-planned, well-balanced program for the year. The subcommittee has gone over all five reports together, and likes them. Due to the lateness of the hour, I suggest that only one of these reports be read. The others are quite similar. I would like to call on Fred Lucas, Chairman of Plainville Community Farm Bureau, to read his program for the year."

PLAINVILLE COMMUNITY FARM BUREAU

Meetings – First Friday of Each Month

JANUARY 7

8:00   Member Participation: **Planning a Business Meeting** – Leader, Clyde Fleenor

8:20   Business Meeting

8:40   Our Farm Bureau Policies: **Report on AFBF National Convention** by our delegate John S. Leigh
Explanation and discussion of Farm Bureau Policies

9:30   Musical numbers – Young Farmers Committee

10:00   Pie and coffee

FEBRUARY 4

6:30   POTLUCK SUPPER

7:30   Member Participation: **Presenting A Good Motion** – Leader, Ernest Hennings

8:00   Members' Night Program: **The Agricultural Outlook, Our Program of Work for the Year** – Fred Lucas, Chairman, Plainville Farm Bureau

9:30   Question and answer period led by Jim Anderson and Fred Lucas

10:00   Refreshments

MARCH 4

8:00   Membership Participation: **Good Committee Meetings** – Leader, Kenwood Krispin

8:20   Business Meeting

8:30   **Rural Zoning in Jefferson County** – Mason McDonald, Member, County Farm Bureau Committee on Rural Zoning
Discussion, questions and answers

9:30   Refreshments

APRIL 1

6:30   BANQUET: Arranged by Women's Committee
Master of Ceremonies, John Benson
Tributes Due – Frank Brand, Regional Director
Musical Selections – The Farm Bureau Quartet of Jefferson County

Farm Bureau's Role in Local Issues – Fred Lucas, Chairman, Plainville Farm Bureau
Dancing and entertainment
Guests: City and County officials and civic leaders

MAY 5

8:00 Membership Participation: Skit: **The Committee will Come to Order**

8:30 Business Meeting
Our Farm Safety Program in Jefferson County
Mrs. Eldridge Roberts, County Chair of Safety, Women's Committee – **Making Highway Safety a Reality**
A discussion with questions and answers, led by Mrs. Merle Climber

Ice Cream and Cake

JUNE 8

8:00 Member Participation: **Streamlining Business Meetings**

8:10 Business Meeting

8:30 **Township Legislative Committee In Charge** – Howard Gilman, County Legislative Chairman
Questions-answered by Panel of County Commissioners

Sandwiches and Coffee

JULY 5

8:00 Member Participation: **Taking Part in Discussion** – Leader: Jonathon Wayne

8:30 Business Meeting

8:40 **Policy Development Highlights** – Ralph Allen, Chairman, County Policy Development Committee
Panel Discussion with questions and answers – Panel: Ralph Allen, Stephen Holden, Frank Gallottee

AUGUST 2

11:00 ANNUAL FARM BUREAU BARBECUE: Stoddard Ranch
Chairs: Vincent and Lola Maroney

SEPTEMBER 6

8:00 Member Participation: **Electing Officers** – Leaders, Caleb and Margaret Caldwell

ANNUAL BUSINESS MEETING: Election and Installation of Officers; Consideration of policy recommendations

Refreshments

OCTOBER 4

6:00 POTLUCK DINNER

COUNTY PROGRAM PLANNING – THE REPORT

7:00   Program: **Silence Is Golden** (a real surprise)
         Young Farmers Committee of Farm Bureau in charge

NOVEMBER 5
6:30   BUFFET SUPPER
7:30   Member Participation: **A Chairman's Memorandum** – Leader, Franklin Bates, Jr.
7:45   Business Meeting
8:00   **Goals and Objectives for the Coming Year** – Herbert Nolan, Chairman, Jefferson County Farm Bureau
         Discussion, questions and answers led by President Nolan
         Guests: County Farm Bureau officers and Board of Directors

DECEMBER 9
8:00   **CHRISTMAS PARTY** at Farm Bureau Hall
         Plainville Farm Bureau wishes you a Very Merry Christmas and A Happy New Year

"Thank you, Fred Lucas," said Chairman Stauffer. "We certainly are proud of Plainville's achievement. The other four community Farm Bureaus have equally interesting and well-balanced reports, as you will see later when they are printed.

"By the way, Abner, you were asked to find out about getting all these Farm Bureau events made up as a calendar. Have you a report for us?"

"Mr. Chairman, I have. Our Service Company and our insurance companies were delighted to make up a calendar of Farm

The Waseca County, Minnesota, membership team take a break during their membership recruitment campaign.

(Photo by Joan Waldoch, American Farm Bureau Federation.)

Bureau events. Quite a space will be devoted to each month and all Farm Bureau meetings will be listed – even the community groups. They're going to do a first-class job on it and they're happy to participate in this project for Farm Bureau."

"Thank you, Abner. It's going to make me mighty proud to hang that calendar up by the telephone."

"Mr. Chairman," said Ken Harding, "I want to express the feelings of all the community Farm Bureau chairmen. We are deeply grateful to you, John Stauffer, and to the rest of the committee for arranging this unified program of work for the year. Each of us has felt somewhat alone and worried about how to do his job well. Now we know we are part of an efficient plan.

"We each have a new understanding of our state president's words – 'the real power of Farm Bureau is in the counties and in the communities.' We are tied into the county powerhouse, and we can and will work as never before. We're mobilized and we're marching straight ahead."

"We appreciate that, Ken," responded John Stauffer. "And I want to say to every member of this committee that you've done a wonderful piece of work. You've done it conscientiously and effectively – like a team pulling together. No one has fallen down on a single assignment – every one has done his part, promptly and well.

"When I make our report for this committee at the annual meeting, I want every one of you to come up and take a seat on the platform while the report is being read. Farm Bureau members should see their representatives who have planned Jefferson County's programs for next year."

"I hope we move in on other things like we have on this program planning job," remarked James Mason. "Planning all the programs for all the groups of Jefferson County looked impossible to me. I counted up, and we had to arrange for almost one hundred meetings. But once we tackled the job right, it was easy. Now it's done."

"Amen," encouraged several voices.

"Mr. Chairman," spoke up Matthew Kitchin, with a grin, "now I understand why my wife was so self-satisfied year before last, when she got all our Christmas presents bought and wrapped by November first. It's a mighty good feeling to have everything planned for the year. Boy – do I feel wonderful!"

"Just one word," chirped McTavish, warningly. "Let's not count our chickens too soon. About all we can do now is just cackle! We have a hatching job for the rest of the year. If everything hatches out as we've planned, then it will be time enough for us to crow."

Over the shouts of laughter which greeted this warning, someone called out, "Mac's still savin' money. He's got our committee doing the hen's job and the rooster's too."

Speakers and Audience

CHAPTER 16

Larry Boys, District Director of the Indiana Farm Bureau, greets members during an evening program at their annual meeting.

(Photo courtesy Indiana Farm Bureau.)

16

Shall We Have a Speaker?

Many times a county or state Farm Bureau will decide to conduct a meeting which includes a speaker.

A well-informed, competent speaker may serve many purposes in Farm Bureau meetings. He can attract members to attend; he can explain and throw light on a puzzling problem; he can clarify facts; often he can answer questions; he can expose weaknesses and strengths in ideas; he can relate proposed policies to the fundamental beliefs of Farm Bureau; he can give voice, and life, and courage to the thinking members; he can stimulate and inspire a group to action.

An outside speaker, or one of your own members, can do some or all of these things, if he has the necessary knowledge and skill. But use speakers judiciously, for they are not the answer to all Farm Bureau meetings. No speaker is the complete answer to any Farm Bureau meeting.

When your committee sits down to plan next year's meetings, don't always take the easy way out. For example, "Now what'll we have for the January meeting? Let's invite Senator Osgood; he'll draw a wonderful crowd!"

Senator Osgood, plus a dessert, will not ensure a successful meeting. True, members may go home saying, "Wasn't the Senator's speech wonderful? and "Did you ever taste such marvelous pie?"

But is that your aim?

Take the surer more constructive way instead. Ask yourselves:

1. What **subject** should the January meeting cover?
2. What is our **aim** for the meeting?
3. What are the best ways to **accomplish** our aim?

When you have determined the subject and the real aim of the meeting, then study your meeting as a whole. Set your sights on what you want to accomplish. Round it out. Launch the meeting in the direction you want it to go.

Consider several ways of reaching your meeting goal. Examine various ideas for using your own members on the program. Allow ample time for some form of discussion. Keep in mind that audience participation is important. Never forget to encourage the folks attending the meeting to get into the act if they want to. Remember that good meetings are often held without any outside speaker.

Having studied your meeting needs thoroughly, if you decide you want a speaker, get a good one. Next, face the fact that even an excellent speaker is only a part of your meeting plan. Your goal is to arouse a sustained interest and usually to secure some constructive action.

When the speaker has informed and inspired the members, you must be prepared to utilize her good work. If you have a speaker who arouses enthusiasm to get new members for Farm Bureau, don't just thank him and bury the enthusiasm by "adjourning for coffee and doughnuts." As soon as she stops, go into action on the wave of her enthusiasm. Call for volunteers to assist the committee, ask for names of prospects, start everybody out to sign up members and report back at a dinner meeting. Inspiration soon dies unless it is followed almost immediately by discussion, decision and action.

SOURCES FOR GOOD SPEAKERS

Farm Bureaus have many sources for obtaining interesting speakers who will draw good audiences. Some sources are:

1. Elected officials of Farm Bureau and staff members are a fine source of speakers. They are prepared to talk on subjects of interest to Farm Bureau members and they know of other good speakers.

2. Farm Bureau's own leadership program, which includes speech making and the conducting of meetings, can often

page 184 SPEAKERS AND AUDIENCE

furnish trained speakers. One way to be sure of having good speakers is to "make 'em." Farm Bureau members enjoy listening to their fellow members who speak well.

3. State, county and city officials welcome an opportunity to speak. They also know when congressmen, senators and national officials may be in the locality.

4. Universities and colleges can usually furnish effective speakers and welcome the opportunity to do so. Many colleges maintain a speakers' bureau; this bureau lists faculty members who are competent speakers and also know of interesting people who will be visiting the campus.

5. Librarians are often able to advise on speakers. They are closely in touch with news, with published developments in agriculture, and with personalities who would be attractive as speakers.

6. Editors of newspapers and directors of radio and television stations often know ahead of time who is coming to a community or state. They know which people are effective speakers. A good speaker with a less impressive title is usually preferable to an impressive title and a poor speaker.

Don't be afraid to "hitch your wagon to a star" by inviting a prominent person. Frequently, well-known personalities are glad to have a chance to talk to a Farm Bureau audience. Sometimes speakers of national importance are looking for an opportunity to talk to a nationwide audience from the platform of a small group. Those whom you invite are complimented by your invitation, even if they cannot accept it.

Arrangements for the Speaker

Only an experienced speaker realizes how directly her talk is affected, for better or for worse, by the attitude and actions of the program chairman, officers and members. Usually a speaker is contributing her time and ability to Farm

Bureau. It is important that she should receive, in return, every consideration from Farm Bureau representatives.

If the program chair and officers are courteous and thoughtful in arranging for the speaker's convenience and comfort, he will be in a happy frame of mind and will respond by giving his best. He will leave with the feeling that he would like to come back again. If, on the other hand, he is forced to shift for himself and to find out facts which should be furnished him, he will be sorry that he accepted the invitation.

Farmers and ranchers are proverbially hospitable, practical people. They know how to entertain graciously and to put visitors at ease. Farm Bureau leaders can make use of these abilities to be good hosts to their speakers or to others taking part in their programs. Since Farm Bureau assumes responsibility for extending every possible courtesy to the speaker, from the time she leaves her home until she returns, there are many questions to be asked and arrangements to be made.

The invitation to the speaker should make clear whether or not his expenses are being paid. If he is coming from a distance, he should be advised as to plane and train arrivals; if he is to stay overnight, he should be told before he leaves home what arrangements have been made for meeting him, what his accommodations will be, and who is to call for him and take him to the meeting.

The person extending the invitation should ask whether the speaker has any special requests: for example, if she is going to show slides, is she bringing her projector with her, or should Farm Bureau secure one? Are there any special arrangements she would like to have made? Would she like to be alone before the program; or would she prefer to visit with some other members?

The speaker will probably want to know some things about the audience as well. What will be the size of the audience? What is the approximate age? What is the male to female ratio? Perhaps you can relate the commodity interest of the members. Will there be agribusiness owners or politicians present? What was the reason for selecting this topic? What is the time allotted for the speaker? Are there other speakers on the program? What type of room will the presentation be made in? Will the seating be around tables or theater style?

When the Speaker Arrives

When the speaker arrives at the meeting place, she may wish to see whether the microphone is adjusted to the proper height and whether there is a convenient place for her notes.

If he has been invited for a luncheon or dinner which precedes the meeting, his host should find out whether he has any special food preferences. Some speakers prefer not to eat before a talk; others want milk or buttermilk instead of coffee. Some prefer to sit quietly; others like to ask questions about the interests and activities of the members. The host should tactfully encourage the speaker to follow his own inclinations. At the conclusion of the luncheon or dinner, and before the program begins, there should be a brief but definite recess during which the speaker and the members may visit a rest room.

Normally, speakers are somewhat nervous before a talk. This is due to the mounting tension which is inevitable if a speaker is to do his best before an audience. Be thoughtful, therefore, of details for his comfort. Tell him where he is to sit on the platform, at what point in the program he will be introduced, and the subjects of the other speakers.

The courteous introducer tells the speaker briefly what she plans to say in introducing her. This gives the speaker time to think of an appropriate reply. It also gives her the opportunity to request, if she wishes, that certain parts of the introduction be omitted or that other facts be added.

Speeches of Introduction are Important

Few people realize that minute for minute, and word for word, a speech of introduction may contribute more to the success and pleasure of an occasion than the main speech itself. Introductions are vitally important to the speaker and to the audience. Yet too often they are either mediocre or frightfully bad.

It is a common belief that anybody can introduce a speaker, that speeches of introduction are unimportant. Actually, it is harder to give an appropriate and effective introduction than to make a longer talk.

Irvin Cobb once said, "An introduction can make or ruin the main speech." A good speech of introduction can launch a speaker like jet propulsion. A poor introduction is like throwing cold water in his face.

How not to Introduce a Speaker

In the hope that all of us may become more sensitive to speeches of introduction, let's consider, in a lighter vein, some of the introducers we could do without.

The first is the **thunder-stealer**. He would never think of lifting the speaker's watch or his wallet, but he is committing a worse crime when he steals something far more valuable to the speaker – his speech.

The thunder-stealer is often an unconscious thief and operates in full view of the audience. "Suppose you tell me the highlights of your speech," he urges before the meeting begins, "so that I can introduce you fittingly."

The speaker complies, and a little later is horrified to hear the introducer repeat his best story or explain to the audience what his climax will be.

SPEAKERS AND AUDIENCE

One speaker thought that he recognized the large white building on the corner as the place where he was to talk. Only after he had hurried in did he discover that he was in a mortuary. As he drove to the proper building in the next block, he chuckled over his beautiful opening story. He planned to tell of his error and to remark slyly that he expected this audience to be a livelier group of listeners than the ones he found at the other place. He confided his joke to the chairman. The chairman made it the high point of his introduction!

If you have heard the speaker before, do not use extracts, or quotes, or ideas from her talk in presenting her. A preview dulls the interest of the audience. No matter how much you know about what a speaker intends to say, do not leave her in an intellectual void by stealing her ideas.

We could do without the **never ender**, who rambles on as though he had swallowed an unabridged dictionary and was regurgitating all the words. He doesn't steal the speaker's speech – he steals his time and the time of the audience too. He seems to fear that he will never get another chance to address an audience and he must make the most of this one.

(Photo courtesy Iowa Farm Bureau.)

Ron Corbett, speaker of the Iowa House of Representatives addresses a Farm Bureau audience.

Peggy Rutherford, Chairwoman of Arkansas Farm Bureau Women's Committee speaks at an Arkansas Farm Bureau leadership conference.

(Photo courtesy Arkansas Farm Bureau.)

On and on he rambles, while the audience grows indignant, and the speaker wilts under the strain. If you are the introducer, don't confuse yourself with the main speaker.

Another irritating introducer is the **apologizer**. "Members and friends" he begins solemnly, "I must apologize tonight for my bad cold. If I had the time, I planned to look up things to make our speaker more interesting, and to prove his intelligence, but I'm sorry I haven't had time. I deeply regret both our poor attendance and the lateness of the hour, and I offer my apologies to our speaker. Lastly, I hope you will forgive me for taking so much of your time. It gives me great pleasure to introduce our last speaker, Mr. Melvin Carruthers, who will talk on the subject of **The History of Tobacco Growing**! Mr. Carruthers."

Apologies are embarrassing and frequently lead the introducer into quicksands of trouble.

"I wish to apologize for the Board of Directors because we failed to get Mr. Getell. You know he is the brilliant legal counsel of our organization, and a speaker whom you all love to hear. But we have done the best we could. We have Mr. Henderson, who is a sort of a lawyer assisting Mr. Getell, I think. Of course, I don't mean for a minute that we aren't delighted to have you, Mr. Henderson, but…"

The **over-builder-upper**, who speaks as though he were the advance agent for a super-colossal television show, is never content to introduce the speaker as a well-known agriculturist" or "the greatest wit of our time."

"Gentlemen, I want to introduce to you the world's greatest humorist. Every time he opens his mouth you will just die laughing. No one can resist his jokes. Before he has talked two minutes, you will find yourselves rolling in the aisles."

By this time the audience is muttering, "Oh yeah? Well, let the so-and-so prove it. I defy him to make **me** laugh!" Instead of giving the speaker a comfortable confidence and arousing the interest of the audience, the exaggerator has built up a wall of resentment which the speaker is often unable to climb over.

Then there is the **clicheist** who uses nothing but outworn and overworked expressions; phrases which were clever the first time they were said, but which have been used so often that they have lost all meaning.

"We have with us this evening a man who is known to all of you – a man who needs no introduction. Unaccustomed as I am to public speaking, I don't know why I was selected for this honor. And so, without further ado, I take great pleasure in presenting to you, Dean Mansfield." The introducer might almost as well have said, "Blah, Blah, Blah!"

Even a slightly new twist can make an old cliche come to life. Walt Disney was introduced at Yale University as the man who "labored like a mountain and brought forth a mouse – with which he conquered the whole world."

Then there is the "Who's Who" type of introduction – "Born in 1920, in Racine, Wisconsin, received his A.B. at the University of Illinois, received his M.A. at Columbia University, served as lieutenant in World War II, author of **Changing World Markets**" – you almost expect the introducer to conclude by saying, "He is survived by his wife, Matilda and two children." Facts are important, but choose facts which are related to these listeners.

We all dislike the **wilted radish** type of introducer who offers no inspiration to the audience and discourages even the most enthusiastic speaker.

"Next on the program," he says dispiritedly, "is Mr. Jim Brown who has something he wants to say to us. I think we can give him fifteen minutes and still finish on time. Mr. Brown."

No wonder Gertrude Stein is said to have mailed this reply to clubs asking what her charges were for a speech: "Five hundred dollars if I am to be introduced. Two hundred and fifty dollars if I am allowed to introduce myself."

Contrast the inadequate or wilted radish type of introduction with the brief but challenging sentence introducing a prominent minister to a New York Farm Bureau: "A man who preaches, teaches, and lives Christianity every day of his life – Weldon Forbes of Nashville, Tennessee."

The prize for annoying introductions goes to the **egotist** – the chairman who talks about himself instead of the speaker. "It gives me great pleasure," he announces with an anything but happy expression, "to present to you a man whom I have known since childhood. I was born in this town, and I grew up in this town. One of my earliest recollections is of stealing apples from the Harding Orchard. My companion in crime is to speak to us tonight. When I was in college…"

"I am happy to present my childhood chum, my college classmate, my good friend, who has so often sought my advice – Senator Newton Evans." The egotist manages to convey the idea that the speaker only shines in the reflected light of the introducer.

Then there is the **digressor**: "I am happy to present to you Mr. Thomas O'Flaherty, who will speak to us on a subject of great interest."

The detour sign goes up, however, with the next words: "But before I tell you all about Mr. O'Flaherty, I want to express my appreciation to the Publicity Committee, who has had a most successful year. I'd also like to ask whether any of you would like to volunteer to serve on the Committee next year. I'd like a show of hands…" By the time this introducer's helter-skelter mind gets back to Mr. O'Flaherty, the speaker has lost the fine edge of his enthusiasm.

The **blunderer** embarrasses everybody. He forgets the speaker's name or his subject, or both. Mr. Harris, who has been invited to speak on **Soil Conservation** becomes "Mr. Ah-ah-um Davis, who is going to speak on **Forest Fires**."

"And, of course, we always keep the best to the last. I know you have been sitting here all evening just waiting to hear our county president," says the blunderer, blandly overlooking the others on the program who have already spoken and who are sitting red-faced behind him.

Good Speeches of Introduction

What is a good speech of introduction? Its real purpose is to build a bridge of interest, confidence and good will between the speaker and the audience; to establish rapport between the two. In general, introductions should be brief; rather a distilled, gem-like characterization of the speaker than a long, factual description.

Sometimes, however, if a speaker is not well-known to the audience, it is advisable to tell them enough facts to arouse their interest in him and in his subject. If a speaker faces a somewhat hostile audience, it may be necessary for the introducer to build a bridge of fair play and willingness to listen to all points of view.

If the speaker is extremely well-known to the audience, it is sometimes better to introduce the audience to the speaker, telling him a little about the group which he is to address. When Marshal Foch visited the United Sates, it was planned to have him stop for one minute to address the wounded veterans in Letterman Hospital, San Francisco. The young veteran introducer started his brief talk, and was startled to feel a hand on his shoulder.

"Pardon me, young sir," said the Marshal. "I have a favor to ask of you. With this bright uniform on, I imagine everybody here knows who I am. I would like to know more about my comrades here in the audience. Will you introduce them to me?"

The young man explained that most of the men belonged to the Rainbow Division and mentioned the areas in which they had served. Then he pointed out three badly wounded men down in front, and in one sentence described the service of each. The Marshal beamed. "Now we have rapport," he said

To the dismay of the many officials who accompanied him and to the delight of the servicemen, the Marshal gave a jolly fifteen-minute talk. Then, spying a big cake down in front of the platform, he asked if he could cut it, and spent another fifteen minutes visiting with the servicemen.

"Best introduction I ever had," he chuckled appreciatively, as he shook hands with his introducer.

There is no one formula for a good introduction. The problem of establishing a warm and friendly interest between audience and speaker is different on every occasion. There are many solutions for this problem. That is what makes good speeches of introduction varied and interesting.

The Washington newspaper correspondents had President Roosevelt as their speaker one evening after his second election. The introducer said, "Mr. President, win, lose, or draw, you have been a newspaperman's president. You have made lots of news and you have served it sizzling hot." The brief introduction aptly summarized the feeling which existed between many newsmen and the President.

Sometimes an introduction creates the desirable climate for both speaker and audience, when the introducer acts as the mouthpiece for the group. For example, if an audience admires and appreciates a speaker, whom they know well, they are grateful when an introducer expresses their sentiments for them. A bridge of love and warmth is built between speakers and listeners.

One president of the American Farm Bureau Federation was introduced to 8,000 delegates at a National Convention. They were delighted and thrilled when, instead of the usual introduction of "Our President," the introducer voiced the thoughts of the members thus:

> "And now we come to the high point in our convention. Our speaker is one of the truly great citizens of our country – the distinguished agricultural leader of the world. A man of great intelligence and ability, a man of unimpeachable integrity, a man of high courage. An Iowa farmer and, I am told on good authority, an expert in the production of

pigs. Our beloved leader, the president of the American Farm Bureau Federation."

Many delegates referred appreciatively to this speech of introduction which gave expression to their individual sentiments. Thus it fitted the situation.

Well-thought-out introductions establish rapport in a few carefully chosen sentences. In introducing a speaker to the Peach Growers Association, one chairman said: "Our speaker tonight is the man who, by pioneering new methods, is giving tree-ripe peaches to the dinner table. He's here to share his experiences with the rest of us – Bradley Holmes."

One commodity chairman presented a professor to the Cattleman's Association thus:

> "Welcome, all 300 of you cattlemen and friends! This big audience is proof of your eagerness to hear the man who probably knows more about brucellosis than any person in the state. Our speaker makes his headquarters at the University Farm, but our herds of cattle are his laboratory. Tonight he's planning to kill two birds with one speech. He will explain the most economical way to handle the mastitis problem, and at the same time he will tell you how to meet the regulations of the state inspection service – Arthur Harding Jones."

In presenting the president of a state Farm Bureau, the introducer said:

> "By electing him president we showed our admiration. By our work this year we proved our faith. Our speaker has just completed an enviable year of achievement, but he avoids the mistake of Lot's wife, and does not look back. That is why he has chosen the subject tonight – The Year Ahead – our state president."

Note how this introducer, although speaking in a light vein, manages to point out all the main achievements of the speaker.

Good introductions take understanding, thought and preparation – but they are rewarding to both speaker and audience. Speeches of introduction are important!

Audience Attitudes

You might be surprised if you knew how closely related you, as a member of the audience, are to a speaker. The average individual in an audience feels comfortably obscure and free to relax and enjoy himself. If he wants to comment to his neighbor, he does. If he is sleepy, he yawns. If he is bored, he fidgets – or maybe leaves. He is unaware of the close relationship between the speaker on the platform and his own behavior as one individual in the audience.

Every member of an audience is most important to a speaker. From her position on the platform she notices every whispered conversation. Are they criticizing her on what she is saying? Worrying, she can't concentrate on her speech. Every roving glance or bored expression disturbs her. Every person who leaves the room wounds her. She is highly sensitive to the behavior of each person in the audience. One discourteous person can ruin a speech.

If you want the best from a speaker, cooperate as a member of his audience. Give him your interested attention. Maybe you're not interested! In this case, all you need to do to fulfill your obligations as a courteous listener is to assume an alert, interested expression, keep your eyes on the speaker, and think about how this subject might affect somebody you know, or how it might change in the future.

If you wish to be an active participating member of the audience, you'll follow the speaker's line of thought, respond to his witticisms, and encourage him by your attitude of appreciation. Even one skilled listener, who understands his importance to a speaker, can stimulate and encourage him to do far better than he might otherwise do.

How Farm Bureau Develops Policies

CHAPTER 17

American Farm Bureau President Dean Kleckner address the convention body at AFBF's seventy-fifth anniversary meeting.

(Photo courtesy American Farm Bureau Federation.)

HOW FARM BUREAU DEVELOPS POLICIES

WHAT ARE POLICIES?

Just as each season brings with it a different type of work to be done on the farm, so Farm Bureau's program of policy development and policy execution has an established cycle.

There is the period of the year for proposing and planning policies – the planting season. There is another period for studying, fact finding and selecting – the growing season. Then there is a final period of putting the agreed-upon ideas into action – the harvest season.

Farm Bureau has an original and highly effective system for developing its plans and carrying out its work. Probably no voluntary organization has a method which assures such effective control by its members or is so practical and efficient. It has been tested by time, and Farm Bureau's achievements over the years are proof of its success.

Farm Bureau operates through policies – developed and carried out by its members. A policy is the statement of an idea, a belief, a method, or a plan. Added together, these policies are Farm Bureau's goals for the year. They may be philosophical statements of ideals, or they may be concrete blueprints for action.

What do policies cover? There may be problems of farmers and ranchers or problems of all citizens. They may be community problems or international problems. They may be goals or methods of reaching them.

What is the aim of policies? Their aim is usually to solve problems of the community, county, state, nation, or even of the world, or they may represent an advance where there is no problem – a new discovery or step ahead.

Are policies concerned with legislation? Policies are directed toward legislation only when members conclude that there is no better recourse or remedy for the problem. Farm Bureau members face a problem squarely, analyze it, and generally work out a solution to the problem themselves. If no other solution can be worked out, then they turn to legislation.

Policies express Farm Bureau's fundamental ideas on community, county, state, or national issues. These problems may be in the field of agriculture – seed certification for example; or in the field of citizenship – the issue of centralized administration of government versus local government. Once determined and adopted, these policies become the plan for the year's work. They are the statement of what a majority of Farm Bureau members have voted, after prolonged study and discussion.

The more basic policies state the philosophy of Farm Bureau – those principles which the members hold fundamental to the welfare of their country and of Farm Bureau. They are the principles of the Christian faith, the Bill of Rights and the Constitution of the United States. These are the yardsticks by which lesser proposed policies are measured.

Farm Bureau's policies form a philosophy of American agriculture, as well as the goals and plan of action for Farm Bureau.

MEMBERS ORIGINATE POLICIES

The members originate and develop Farm Bureau policies. Often a need gives birth to a policy; several farmers in a community feel strongly that "something must be done" to correct a situation, to solve a problem, or to achieve some new advance in agriculture or government. The subject may be one of community concern – a new road, a change in the assessment rate, a marketing difficulty, or a new zoning ordinance. Often, however, the farmer looks beyond his fence rows to the needs of agriculture, the country, or even the world. He is a citizen as well as a farmer and is concerned with all of the problems of a citizen.

Introducing Recommendations

Usually policy recommendations are first presented at a neighborhood or community Farm Bureau meeting on policy development. You or some of your friends get an idea which you think would benefit your community, your state, or your country.

If you have only the germ of an idea which is not yet formulated, you probably discuss it with some of your friends and neighbors. During discussion the idea begins to crystallize. Talking over your idea with others will help you to eliminate the bugs. Then you are in a better position to prepare a good recommendation for a policy.

Writing a recommendation is not easy, nor is it impossible. Your first job is to collect the facts. You investigate, study and seek advice from qualified persons. Many Farm Bureaus have study committees which assemble and weigh facts concerning proposals. Before a recommendation is approved, you, as its sponsor, are likely to be asked many questions about it. You will need plenty of facts, figures and proof. So tap the sound ideas of other members.

You will want to word your recommendations carefully. You don't use "whereas's" or "wherefore's" or big words. You do need a sound idea, and you must express that idea clearly. Here is a policy of the American Farm Bureau Federation covering a vital national issue. Notice how clearly and simply it is stated, and how brief it is.

> SAFETY: We recommend that state Farm Bureaus cooperate fully in sponsoring the enactment of uniform traffic and vehicle safety laws and use their influence to assure strict enforcement of speed laws. We encourage the development of driver education courses in our schools.

One county Farm Bureau, in a dairy area, was delighted when its proposed policy recommendation on promoting dairy products was adopted at the Annual Meeting of the American Farm Bureau Federation. Some of the members pointed out proudly that the wording of the adopted national policy was almost identical with the recommendation which they had approved at their county Farm Bureau meeting. This was a tribute to those who drew up the recommendation.

The member who blithely tosses a recommendation into the ring at the policy development meeting of his local Farm Bureau, with some such comment as, "I don't know too much about this, but we might as well give it a try," is wasting his time and that of his fellow members. In competition with well thought out, sound, and carefully worded recommendations, his is the weakling of the litter and will probably die aborning.

Before you present your idea, try to think it through. Test it to see whether it harmonizes with the fundamental beliefs of Farm Bureau. Recommendations, if they are to become policies, must be well-phrased, basically sound and supported by information and facts.

Policies and Community Farm Bureaus

When the recommendation for a policy has been carefully prepared, it is usually introduced as a motion at a meeting of a local Farm Bureau group for discussion and vote. Sometimes during the discussion it is amended and reworded. Here in the neighborhood and community group is the place to try out arguments, discover weak points, test member reactions and reach agreement.

If your recommendation is approved, and if it concerns a matter of community interest only, it becomes an adopted policy of your community Farm Bureau. If it covers a matter which is of county, state, or national concern, it immediately becomes a recommendation of your community Farm Bureau and is forwarded to the county resolutions committee.

Policies and County Farm Bureaus

Here the recommendation faces its second test. The county resolutions committee studies it, searches for more facts, and compares it with policies already adopted and with similar recommendations which may have been sent in by other community groups. The committee may combine recommendations which differ only in minor points. It may reword the recommendation or make changes in it. The committee frequently holds hearings on the recommendations. It often consults with well-informed persons and asks advice of technical people.

If the county resolutions committee approves the recommendation, they present it to the county annual meeting as a policy recommendation.

The Farm Bureau members at the county annual meeting may amend the recommendation during discussion. If they adopt it, and if it covers a subject which is of concern to the county only – such as a zoning ordinance or a change in the county tax rate – it then becomes an adopted policy.

Usually the county Farm Bureau, at the conclusion of the county annual meeting, provides the press with copies of the actions taken. These are in two groups. The first group lists the actions taken with respect to local issues. These are not the official policies of the county Farm Bureau. The second group lists the county's recommendations concerning state,

Rex Lukow of Adams County Farm Bureau in Nebraska speaks to fellow voting delegates about a policy issue during the state convention.

(Photo courtesy Nebraska Farm Bureau.)

national, or international issues. These are now recommendations to be forwarded to the state resolutions committee.

Policies and State Farm Bureaus

This committee is usually appointed by the state president in consultation with the board of directors, and its chairman is frequently the state vice-president. The committee undertakes more study, investigates and compares all policy recommendations which may have been received from the counties. The resolutions committee may amend or clarify proposed recommendations and may combine the best features of several similar recommendations into one. More facts and statistics are sought, and often hearings are held. If similar recommendations are made by a number of counties, this indicates wide backing for the idea and the chances for approval are good.

If the state resolutions committee approves the recommendation, it is forwarded as a recommendation to the house of delegates of the state Farm Bureau. If the recommendation concerns only this one state, it becomes an adopted policy of the state Farm Bureau when approved by the house of delegates. It is then entitled to the support of all county and community Farm Bureaus in the state.

Policies and American Farm Bureau Federation

If, however, the recommendation is of concern to several states, or to the whole country, it is forwarded as a recommendation to the Resolutions Committee of the American Farm Bureau Federation. This committee is a thoroughly representative group. The chairman is usually the vice-president of the American Farm Bureau Federation. Its members include the elected president of each member state Farm Bureau, the chairman, vice-chairman, the chair of the

HOW FARM BUREAU DEVELOPS POLICIES

Women's Committee and the chair of the Young Farmers and Ranchers Committee.

The policy recommendations which come to the Resolutions Committee of the American Farm Bureau Federation are carefully scrutinized and screened. Those which are similar, yet overlapping, are combined. Those which conflict only slightly are harmonized. Their legal implications are analyzed. Their wording is revised. Their meaning is carefully tested. Those which are approved by the Resolutions Committee go to the voting delegates as recommendations.

Those recommendations which receive a majority vote of the voting delegates become Farm Bureau policies with regard to national and international issues, to be supported by every Farm Bureau member. These policies, together with those retained from other years, make up the Farm Bureau policies and program of work for the year. They continue in effect unless they are modified by the voting delegates of the member states at a future annual meeting. They are the policies covering national issues for all Farm Bureau members everywhere. They are the established objectives of Farm Bureau.

The annual meeting of the American Farm Bureau Federation thus climaxes the policy development process in which hundreds of thousands of farm folks in neighborhood, community, county and state meetings throughout the nation have participated.

John Phipps, Edgar County Farm Bureau, Illinois makes his point during a policy discussion at his state annual meeting.

(Photo courtesy Illinois Farm Bureau.)

One Community Develops a Policy

The coroner cleared his throat. "Members of the jury," he began, "you are assembled today for the purpose of hearing the facts in the deaths of Mrs. William Conrad, age 31, and Marsha Ann Conrad, age 4. It is your duty to determine the cause of death and to fix the blame where it should be placed. On August 14th the bodies of Mrs. Conrad and her little daughter were found beneath their overturned automobile at the intersection of Thompson Road and Oak Creek Road.

"The first witness is Jack Engel. Mr. Engle, you have been sworn and are under oath; will you tell your story?"

"Can I tell it in my own words?" queried Jack Engel.

"You just go ahead, and if you get off the track, I'll stop you," said the coroner in an encouraging tone.

"Well – it was this way. I was driving my truck to market with a load of potatoes. Just as I got near the corner where Oak Creek Road crosses Thompson Road, I heard an awful crash. I saw one car roll over twice and the other shoot off into the cornfield. As you folks know, all four of those corners have high corn growing right up to the edge of the road.

"There were three servicemen in the car that landed in the cornfield. They jumped out and we all tried to pry open one of the doors of the other car. I could see it was Jane Conrad and Marsha Ann, inside. You could tell that Marsha Ann was dead – it was horrible – we couldn't be sure about her mother. Finally we got a door open and she seemed to be dead, too.

"The sergeant offered to drive down to the nearest house and phone for an ambulance, but I wouldn't let him because I thought he was too upset to drive. He kept saying, 'I couldn't see anything for the corn – I couldn't see anything for the corn – my own little girl's just about the same age.'

"So I left the servicemen with the bodies and drove my truck to the Anderson farmhouse. I called the county ambulance

HOW FARM BUREAU DEVELOPS POLICIES

and then I tried to call Bill Conrad, but he must have been working out on the farm somewhere because there was no answer. Then John Anderson said we should call the sheriff and he did it. We took some sheets to cover the bodies and John drove back to the wreck with me."

"Are there any further comments?" asked the Chairman of Stony Ford Farm Bureau.

"Mr. Chairman," said Harvey Wood, getting up with grim determination, "I was on the jury at the coroners inquest when the Conrad case was heard. We owe it to Bill to do something. Something's got to be done about these cornfields that make corners blind. I move that we discuss this problem informally."

After the members approved the motion, Harvey Wood continued.

"This is the eighteenth death in this state this year caused by high corn blocking the view at corners. Farmers and people who live around here know the danger, but people from the city and other places don't realize the hazard. I think we should do something about this now. I think if we pass a resolution asking each farmer in this community either to cut down his corn for 100 to 125 feet back from the corners, or to plant a low crop like soy beans in that space, most farmers would cooperate."

"If we could eliminate those blind corners here in this community, I am sure that the other Farm Bureau neighborhood and community groups would join with us in getting the same recommendation through the county Resolutions Committee and the county Farm Bureau meeting. If we do that, I'd be willing to bet that with the help of the county Farm Bureau, we could get this to our state Farm Bureau Board and make this movement statewide. Through our state Farm Bureau we could work with the state Safety Council and the Automobile Association. What do the rest of you think?"

After discussion the meeting continued. "Now this convention will hear the reports of standing committees," announced state Farm Bureau President Hammond. "Mr. Hemphill, will you report for the Safety Committee?"

"Mr. President and delegates," said Mr. Hemphill, "as Chairman of the Committee on Safety, I have been asked by our state president to announce a special award. Mr. Harvey Wood, will you please come to the platform as representative of Jones County Farm Bureau?

"Sixteen months ago, Mr. Wood, you introduced a recommendation at the meeting of Stony Ford Farm Bureau. It called for the voluntary cooperation of the farmers in your community in eliminating a hazard which has caused many deaths. Roads running through corn areas had blind intersections because the drivers' view was cut off by 8 or 9 foot-high corn planted right up to each corner.

"As Chairman of your Stony Ford Committee on Safety, you gained the cooperation of every farmer in your community. This spring, a triangle of land on every corner was planted to low crops for a distance of at least 100 feet back from the corners. Last year there were four deaths, in your county alone, caused by this corn-planting hazard. This year there were none.

"You and your committee did not stop when the problem was solved in your community. You sent a policy recommendation to your county Resolutions Committee. You spoke effectively on it at your county annual meeting and enlisted the enthusiastic support of Jones County Farm Bureau members. Your county Safety Committee worked aggressively as the recommendation was forwarded to the state Resolutions Committee. When the state convention of Farm Bureau approved it, you spent untold hours working in cooperation with the Automobile Association and the state Safety Council.

"You sowed constructive seeds and you have harvested a bumper crop. The accidents in rural communities of this state, caused by blind intersections due to tall corn, have been virtually eliminated. You, your committee and your

17

county Farm Bureau have furnished good leadership, have saved many lives and have performed a genuine service for the citizens of this state.

"On behalf of Governor Hodgkins, on behalf of the state Safety Council, and on behalf of our state Farm Bureau, I present you, as the chosen representative of Jones County, with this plaque, which is a token of expression of our admiration and gratitude.

"Members of Farm Bureau, here is a splendid example of a community Farm Bureau solving a community problem. It is also an example of how one community can spark a whole state to solve its problem. Farm Bureau members in their community groups all over this country are studying and solving problems."

Members Decide Policies

Final decision on every policy is made by the Farm Bureau group – community, county, state, or national – whichever it primarily concerns.

A recommendation affecting only a single county is decided on a county basis by the county Farm Bureau members. If it is approved, it is supported by all county Farm Bureau members as an adopted county policy.

A recommendation affecting only a single state is decided on a state basis by the directly elected representatives of county Farm Bureaus.

A recommendation affecting the nation is decided on a national basis by the elected voting delegates of the member state Farm Bureaus. The voting delegates are the elected representatives of every Farm Bureau member in the country. Thus an issue decided by the annual meeting of the American Farm Bureau Federation merits the support of all member state Farm Bureaus and of all members everywhere. Each member has had a voice and a vote in deciding every policy

HOW FARM BUREAU DEVELOPS POLICIES

of Farm Bureau.

When the American Farm Bureau Federation has decided its policies for the year, these are printed and made available to every Farm Bureau member. This booklet of Farm Bureau policies is the organization's goal and guide for the year. Each state Farm Bureau also prints copies of its adopted policies for its members.

This highly efficient process of presenting policy recommendations in community and neighborhood meetings, of testing them through successive steps where they are studied and improved, and of adopting, finally, the best of them, is unique to Farm Bureau. It is sound in action and trustworthy in results.

The policy development process is the core of the purpose of the organization – to determine what farmers and ranchers are for and what they are against. It is also the core of Farm Bureau's method of work.

This process charts a clear course and well-defined objective for the year ahead. It assures members that Farm Bureau will march forward toward well-considered, carefully chosen objectives, outlined by policies which the members themselves have initiated, studied and decided. There is no more effective process for a program of work. It is true self-government.

Florida state Young Farmers and Ranchers Committee meet with Commissioner of Agriculture Bob Crawford.

(Photo courtesy Florida Farm Bureau.)

CARRYING OUT FARM BUREAU POLICY

CHAPTER 18

California members encouraged adoption of Farm Bureau policy during a visit with their congressman. Pictured are Phil Larson, Craig Pedersen, Congressman Bill Thomas and Ben Laverty.

TYPES OF FARM BUREAU POLICIES

When a Farm Bureau policy has been adopted, the process of carrying it out begins immediately. Most policies are concerned with things that Farm Bureau members can do for themselves through their local groups.

The carrying out of such policies may mean widely varied activities. Examples of policies are: promoting a farm safety program; encouraging Farm Bureau members to take a more active interest in some local school problem; helping to formulate a better safety program for highways; developing a plan to eliminate a livestock disease, such as brucellosis; conducting an educational program to encourage the production of improved farm products, such as meat-type hogs; arranging for young farmer trainees from other countries to spend a year with Farm Bureau members on their farms; or organizing a cooperative purchasing or marketing service.

Policies of this type are designed to stimulate constructive local activities and to meet local problems. They are policies which local Farm Bureaus develop and put into action themselves.

In situations where a community, a county, or a state cannot solve a problem or make an advance by its own efforts, recourse to legislation may be necessary. The carrying out of a legislative policy may mean getting a new law passed, changing an existing law, or preventing the passage of an undesirable law.

Community and county Farm Bureaus constantly face problems which they themselves must solve. In order to solve a problem, they must develop and approve a policy and work out plans for carrying it out. The way in which local groups carry our policies can be more clearly explained by typical examples than by discussion of methods.

A Community Solves a Transportation Problem

The farmers in one wheat-raising community were frustrated and angry because the railroad failed to cooperate with local growers. The wheat crop was the largest in ten years. The storage space on farms and in local elevators was exhausted. Yet newly harvested grain waited for weeks at loading points and was often damaged by weather. Freight cars promised for a certain day did not arrive. Days later they came but, when loaded, they were ignored for several weeks. Letters of protest to the railroad brought no satisfaction.

A committee from the community Farm Bureau visited the station agent in a nearby town and explained their problem. "I don't have any say about the freight cars – I just work here," was the only reply to the committee of frustrated farmers. Next evening the report of the committee's visit was given at the community Farm Bureau meeting.

A member offered this resolution:

"I move that we try to secure better cooperation from the railroad in moving our wheat.

"I believe our committee should take this to our county Farm Bureau Board so that it may take further action in this matter. If necessary, this problem should be taken to the state traffic manager of the railroad, through our state office."

When discussion was called for, one speaker suggested that "our committee make a survey of the whole problem." The committee was instructed to get definite figures on the amount of wheat still to be shipped, the time that it should be loaded, and the number of cars that would be needed so that all of these facts could be given to the county Farm Bureau Board.

A survey was made and the facts were presented in writing to the county Farm Bureau Board. This board contacted the state Farm Bureau office and a conference was arranged with the state traffic manager of the railroad. As the committee was leaving, he shook hands cordially. "I can't tell you how grateful I am for calling our failure to my attention. The facts which you have presented are just what we need to straighten out the whole mess. Instead of complaints only, you have brought me constructive, practical facts which I can use to cooperate with you. If I can get this information regularly, about three weeks ahead of time, I can almost guarantee prompt service.

"We need the business and you need the service. I think you will be as happy in the future as you have been disappointed in the past. If we fail to send sufficient cars, to send them on time, or to ship them out promptly, please call me collect, and I'll do my best to see where the trouble is."

A COMMUNITY CARRIES OUT A LOCAL PRODUCTS POLICY

Let's drop in for a few minutes on two meetings with the Fruitvale Community Farm Bureau and see how they carry out one policy.

MEETING OF MAY 15

"The chair recognizes August Diefenbach."

"Mr. Chairman, we have a problem here in Fruitvale. I move that Fruitvale Farm Bureau adopt a policy of attempting to see that the chain stores in town buy fresh fruits and vegetables locally, and that a committee be appointed to carry out this policy."

MEETING OF JUNE 11

"And now we come to the reports of special committees. Frank Anderson, will you report for the special committee on promoting the sale of local produce?"

"Mr. Chairman. On May 15 Fruitvale Farm Bureau adopted this policy: 'Fruitvale Farm Bureau believes that our local chain stores should buy local produce instead of shipping in fruit and vegetables from New Lockland.' A committee consisting of Frank Anderson, Chairman Elizabeth Kranz and Martin Elsworth was appointed to carry out the policy.

"The Committee met at my home on May 18th and made plans. Next day all three of us went together to call on the chain store managers. Both managers listened courteously while we explained the golden rule that if the stores wanted our patronage, they should give us their patronage. Both managers explained that the reason why the produce was shipped in was that the purchasing agent for the chain stores could buy cheaper in quantity lots, even though there was an extra cost for shipment. Both managers agreed to allow us to try to prove to them that they could make more money by buying locally.

"On Monday, Tuesday and Wednesday of the following week, our committee, assisted by twelve additional members of the Women's Committee of Farm Bureau, gathered produce from our members and delivered it to the stores. The Women's Committee prepared the displays of fruits and vegetables.

The buying public knew nothing of the experiment. The store managers kept track of sales and checked these against the costs which normally would have been paid to the local farmers.

"The joint conclusion of our committee and of the two managers is:

1. That the local produce, furnished by Farm Bureau members for the three days, cost 42 percent less than would the same quantity of produce purchased in New Lockland and shipped in.
2. That the total profits on the produce for the three trial days were 63 percent higher than for the same three days of the preceding week.
3. That the displays of fruits and vegetables had much greater sales appeal, due to the quality and freshness.
4. That the sales managers would present a full report of Farm Bureau's demonstration to their sales department

and would endeavor to obtain permission to purchase substantially all produce from local growers.

"This concludes my report, but I would like to add a few comments. The women of Farm Bureau did a professional job of preparing produce and displaying it.

"I might add that the assistant sales manager of the Ketchum's Grocery Store came in late on Tuesday afternoon, looked over the display and the figures, and told the store manager, in the presence of Elizabeth Kranz, that from now on the Fruitvale store would buy most of its produce locally. He complimented Farm Bureau on its practical approach in achieving its objective. He intends to write a letter expressing the gratitude of the Ketchum's Grocery Store to Farm Bureau. We are assured informally of a similar response from the other store.

   Elizabeth Kranz
   Martin Elsworth
   Frank Anderson, Chairman"

A County Carries Out A Road Policy

The report of the Riverwood County Road Committee of Farm Bureau tells a proud story of achievement by a county Farm Bureau. It is typical of the practical policies which are constantly being carried out all over the country.

Report of Riverwood County Road Committee

"On September 16th this committee was assigned the job of putting into action a policy adopted by Riverwood County Farm Bureau. It read, "We believe that the county road known as Regan Road is vital to this county as a farm-to-market road. We believe that the road should be resurveyed to eliminate sharp curves, should parallel the river for most of its length, and should be paved."

"On September 20th the special committee to put this policy into action met at the county Farm Bureau headquarters and drew up the following plans:

PLANS FOR REGAN ROAD IMPROVEMENT

1. To present a letter stating and explaining Farm Bureau's adopted policy at the meeting of the Board of Supervisors on October 1st.

2. Committee members are assigned work as follows:

    A. Everett Oaks and Molly Bergen are to call on the county engineer, explain Farm Bureau's policy to him, and seek his advice and cooperation.

    B. Willard Peacemaker and Jonathan Treadwell are to dig out facts and figures, particularly figures presenting a comparison between the cost of doing a thorough job of surveying, rerouting, and paving Regan Road as opposed to the cost of doing a little regrading, smoothing out bumps and adding gravel.

    C. Frank Jordan and Bedford Ames are to prepare plans for getting out the facts and information to the supervisors, members of Farm Bureau and the public, by publicity in the Clario and broadcasts from station FBAC.

    D. Bill Johnson is to make a survey of the number of farmers depending on Regan Road for access to town.

"On October 8th the Board of Supervisors invited all citizens interested in Regan Road to attend their meeting. One hundred seventeen Farm Bureau members and 46 other citizens attended, and 14 spoke.

"After a hearing of three hours, the Board of Supervisors voted to resurvey Regan Road, eliminate difficult curves and make it a paved highway.

> Jonathon Anglo
> Herb Edwards
> Alec Wilson, Chairman"

Steps in Carrying Out a Policy

The process of carrying out a Farm Bureau policy has definite steps. **The first step is planning.** What information is required? Where can it be obtained? What type of committees will be most effective? Is local financing involved? How is the program to be promoted? Is the cooperation of other groups necessary? What publicity is needed? What should be the timetable?

Can the problem be solved without legislation? If legislation is required, by whom should it be enacted? How is the particular policy to become a law? Who is to write the bill? What is the timing on introducing the bill? Which legislators will support it? What support from the membership will be needed?

Planning includes the answers to these and scores of other questions.

Planning also includes agreement as to how support for a policy is to be organized both inside and outside of Farm Bureau. What is to be done? How? Who is to do it? When? If possible, several alternative plans are worked out so that if one bogs down another can be tried.

The second step is securing widespread understanding of the policy and the need for action to put it into effect. Everybody must understand what the policy means, what it is intended to accomplish and how these results are to be achieved – members of Farm Bureau, staff members, officers and committees, friends of Farm Bureau, legislators, the people.

The primary objective in policy execution is to get support for Farm Bureau policies. This can only be done through a thorough and widespread understanding of:

1. The need for policy
2. How the policy will meet the need

When Farm Bureau has adopted a policy it must "carry the gospel" everywhere – to the members, businessmen, laborers, professional men, legislators. An informed public is Farm Bureau's goal. Understanding and goodwill put to work, are the most powerful practical forces in the world for getting things done.

When planning is complete and understanding is widespread, the foundation has been laid for **the third step – action**. Key people and key committees in every community know their jobs and are at their stations. Plans for action have been explained and all preparations made, so that when signals come requesting action or support, the response will be strong and instantaneous.

If the policy is one requiring national legislation, Farm Bureau members all over the country swing into coordinated action. It is this synchronized response which furnishes the tremendous power that translates Farm Bureau policies into reality. The steady, loyal, two-way cooperation between the individual member in the community and the national legislative leaders, is the motive power which puts Farm Bureau policies into action.

Policies which do not involve legislation require the same careful planning and coordinated effort of Farm Bureau members in the county or community if the problem is local, in the state for state-wide problems and across the nation for national and international problems.

New York county Farm Bureau leaders meet with state law makers to promote the adoption of Farm Bureau policy.

(Photo courtesy New York Farm Bureau.)

18

How Some Policies Become Laws

Certain of the policies adopted by the voting delegates at the convention of the American Farm Bureau Federation deal with problems and issues that affect the welfare of all farm people and of the country as a whole. On some of these, Farm Bureau hopes to get legislative action. Each newly adopted policy has come through strenuous steps, but once it starts on the path which may lead to the enactment of a desired law, or to the defeat of an objectionable one, it faces the hardest test of all.

The process of securing the enactment of a law by Congress is complex, technical and difficult. Few citizens have even a hazy idea of the strategy, patience, perseverance and skill required to prepare and introduce a piece of legislation, shepherd it through the maze of Senate and House committees, bring it to a favorable vote in both houses of Congress, and finally secure the signature of the President of the United States on it. Enactment of legislation in the states is likewise difficult, technical and complicated.

Fortunately, Farm Bureau has recognized the need for highly skilled and experienced men who take the adopted policy, help to write it into a bill, find sponsors for it, and start the proposed legislation on its way toward enactment as a law.

Farm Bureau's representatives in Washington or the state capitals spearhead the effort on legislation. But it is the members themselves who supply the force which enacts the policy into legislation. Farm Bureau's representatives are the on-the-spot advisers, the eyes and ears of the members, the aides who help the members do the legislative job for themselves.

Farm Bureau's Voice in Washington

Our Federal government, including the President, members of the Cabinet and the members of Congress, leans heavily upon the advice of those who represent the able and conscientious voluntary groups of this country. Farm Bureau is recognized as an organization of integrity, loyalty and strength. Farm Bureau's representatives are accepted as qualified, dependable, patriotic advisers and are consulted frequently.

National officials and legislators of both parties seek and follow Farm Bureau's guidance in agricultural matters. They know the conviction of Farm Bureau's members – that what is good for America is good for agriculture – that farmers can be free and prosperous only in a country which is both prosperous and free.

Farm Bureau's reputation for integrity in legislative matters has been carefully upheld. Members of some organizations have small chance, or no chance at all, to voice their opinions on proposed legislation. Legislators know this and do not take seriously the representative of an organization who says, "Gentlemen, I speak for 50,000 members," but who presents no proof of his representation nor of the opinions of the 50,000.

When Farm Bureau's legislative representatives state that Farm Bureau members favor or oppose a measure, legislators know that these representatives speak from knowledge and that they have proof. They know that the decisions of Farm Bureau's hundreds of thousands of members represent the consensus of opinion of American agriculture and are based on extensive discussion and intensive study. They also know that the statements of Farm Bureau's representatives are backed up by the recorded votes of the individual members in each community, county and state.

Response from the Members

When legislation is pending before Congress or a state legislature, the representatives of Farm Bureau often call for the help of state, county and local units, and of individual members – and they need it quickly. Every member can help by urging his senators and his congressmen to vote and work for a law which Farm Bureau favors, or against a measure which Farm Bureau is opposing.

When an immediate response is needed on a legislative issue, telephone calls, telegrams and airmail letters go to legislators. If the desired response is a long-term one, state offices contact all of their county Farm Bureaus which, in turn, call county, community and neighborhood meetings, make detailed plans, and forward the results of their work to the state offices.

Sometimes the united response from the state Farm Bureaus is enough to convince Congress that the proposed legislation should pass, or should be defeated. At other times the direct help of the individual members is needed.

Carrying out a policy involves not only response to the many requests and suggestions from national and state leaders, but also intense activity initiated by the county and community Farm Bureaus and by the members themselves. National Farm Bureau offices in Chicago and Washington, state Farm Bureau offices, county offices, boards of directors, committees, staff members and all individual members of Farm Bureau, fulfill their particular responsibilities in carrying out the plans.

Meetings are called, conferences are held, boards convene, telegrams are dispatched, radio talks explain, television programs demonstrate, magazine and newspaper articles clarify and publicize, pamphlets, bulletins, speakers, all combine to supply information to members, to legislators and to the public. Farm Bureau offices throughout the country hum with these activities, and members work long hours. To be done properly, all of this supporting work requires time and can't be put off until the last minute.

State, county and community officers and committees are in constant readiness to alert their members to back up their legislative representatives. Telephone committees are ready to reach thousands of Farm Bureau local leaders in a short time. The Women's Committees of Farm Bureau often undertake this task and do telephoning and letter writing. Every member needs to keep thoroughly informed on legislative matters which are pending before Congress or his state legislature so that he can respond intelligently and speedily when legislative leaders call for support.

Know Your Legislators

How can you prepare yourselves to give a quick and intelligent response to a call for help from your legislative representatives?

You can learn the names and voting records of your legislators and often meet them personally. If you know your congressmen, your senators, the members of the state legislature, the county supervisors and the town council members, it will be easier, not only to write to them or to see them yourself, but to persuade your friends to write or call also.

It is not enough just to know a legislator's name and address – you should know him. If you do not know him personally, you can know him by his record. Through Farm Bureau publications, radio and television programs and newspapers you can watch the votes of your legislative representatives. Find out and remember how they voted on legislation in which you are interested.

Respond Effectively

Be alert to answer a call to assist in legislative decisions. Write to your congressman if he is in Washington. Telephone him if time is important. If he is at home, get him on the telephone or, better still, drop in to see him. Let him know how you and your friends feel about pending legislation.

Remember this – since his authority is given to him by the voters of his own district or state, each legislator is interested primarily in his own constituents – that is, voting residents of his district. If you live in California, it may be helpful to write the senator from Missouri, but you cannot hope to gain the response from him that you will from your own legislators whom you help elect. The people who live and work in a legislator's district or state are Farm Bureau's surest means of securing his vote. Farm Bureau's influence on legislation stems from the local communities.

When you write your legislator, be sure to tell him that you are a voter in his district.

> Dear Congressman Jones,
>
> As one of your constituents, I urge...

Then tell him what bill you want him to support or oppose and why. Any legislator is interested in the desires of a voter from his district.

Do not fill out a form or copy somebody else's letter; write in your own individual way. Remember that legislators are just people and don't like form letters any better than you do. For days, one congressman from the Midwest carried in his pocket a letter from one of his constituents and showed it proudly. The voter had penciled his message on a piece of cardboard torn from a placard on a freight car where he helped load potatoes.

By-Products of Farm Bureau's Policy Process

Farm Bureau's system of policy development and policy execution is carefully tailored to the needs of its members. Instead of borrowing the work plan of other organizations, Farm Bureau has created its own, and it is excellently suited to its members. It fits farmers' thinking and attracts outstanding farmers in every community.

The end product of the process is, of course, the policies – both those which are achieved and those which are goals to be sought. Its by-products, though less obvious, are invaluable.

In the first place, the process determines clearly what Farm Bureau members are for and what they are against. Thus it establishes sharply defined goals and a definite program of work.

The policy process also captures the interest and enlists the participation of all members because they themselves propose the recommendations, discuss them, decide them and help carry them out. Members who own and run their organization themselves cannot help being interested members.

The system ensures cohesion all through Farm Bureau; the constant, direct, two-way flow of communication between the American Farm Bureau Federation and the state Farm Bureaus, between each state and its county Farm Bureaus, and between each county Farm Bureau and its members, knits the organization into a unified whole.

Under Farm Bureau's policy system, each member wins a scholarship providing a lifelong, diversified education. Farmers tackle and wrestle with agricultural problems and also with the multitude of issues which challenge every citizen. Gaining an understanding of these problems is an education in itself. The process stimulates members to search out all possible information on a problem. Farm Bureau encourages its members to look for reasons and figures and to consider all sides of a question. Members then apply their best judgment to the facts.

Farm Bureau is acutely aware that the judgment of a group is no better than the facts on which the group judgment is based. It is vital, therefore, that all possible related information both for and against a proposed policy, be examined and the consequences of different suggestions be explored.

Farm Bureau provides facts in many forms. You can't escape them. Members, staff, speakers, magazines, pamphlets, radio, television, meetings, conferences – all combine to marshal the facts. The result of all this fact-finding is that Farm Bureau members are citizens with well-rounded education.

Moreover, every member is thoroughly educated in the principles, methods, aims and activities of his organization. Stop a Farm Bureau member anywhere and ask him what Farm Bureau's policies are for the year – but don't do it unless you are prepared to listen for an hour or so, because he knows and will tell you. Usually he will also explain why Farm Bureau is for these policies and what progress it is making in carrying them out.

The policy process provides constant opportunities for training and experience in discussion – an ability useful not only in Farm Bureau, but in any organization. Neighborhood meetings, community meetings, county and state meetings, committees, conferences and boards all draw members into extensive discussion. In the give-and-take of discussion, members learn to uncover the truth and pick out the fundamentals. They gain the ability to strip from problems the prejudice, uncertainty and misinformation which surround them, until they have reached the hard core of truth. This process of getting at the truth requires keen, straight thinking and develops members' reasoning power and judgment.

Farm Bureau and Individual Freedom

CHAPTER 19

Indiana Farm Bureau members are encouraged to support the candidate of their choice at a county Farm Bureau office.

(Photo courtesy Indiana Farm Bureau.)

The Role of Farm Bureau

Probably Farm Bureau's greatest contribution to this country has been its vital role in support of freedom of the individual and of a government designed to safeguard this freedom. Farm Bureau has fought long and effectively in support of individual freedom and against centralization of dictatorial power in government. Its members have struggled successfully against the many legislative proposals which sought to encroach on individual liberty.

Adherence to the principles on which our forefathers founded this nation and the upholding of the government which safeguards those principles, requires not only eternal vigilance – it requires positive progress around the clock. There can be no slackening, no stopping to coast on past achievements, no long vacations for congratulations.

Steady progress and continuous achievement in seeing and solving new problems is the price which must be paid for the maintenance of freedom. "Only a persistent, positive translation of the faith of a free society into the convictions and habits and actions of a community is the ultimate reliance against unabated temptations to fetter the human spirit."

Farm Bureau's history is one of prolonged and effective struggle against action which would weaken and for action which would strengthen its fundamental philosophy of freedom of the individual. Farm Bureau's part in that struggle may be read not only in national and world affairs, but even more clearly in the record of almost every community, county and state.

One important aspect of individual freedom is the maintenance of a climate where each citizen is free to labor, save, invest and bequeath the results of his savings to his children.

Here is one instance where Farm Bureau recognized a serious threat to individual freedom, developed a policy of opposition and battled almost alone against an old socialistic threat and for the welfare of all the people of a state.

It is told as a dramatized story. But the essence of Farm Bureau's many struggles in defense of individual freedom is distilled in this story.

Farm Bureau's Fight for Freedom

Scene I

The year is 1950. Three farm families in the hard spring wheat area are enjoying dinner together at the home of the Svensons. The guests are the Olsen and the Ericson families. The Svenson family includes Sven Svenson, now 72 years old, his son Carl Svenson, a farmer minister, his wife Helen and their children, Carl Junior and Mary.

"Let's leave the dishes for awhile and go hear the eight o'clock news," suggested Helen Svenson to the group sitting around the big oak dining table.

"Gosh, Mum, that's a wonderful idea, " said Mary, bolting her last bite of apple pie.

"We'll all help with the dishes, anyway, after the news," agreed Hilda Olsen. "Let's take our coffee cups in with us – I'll bring the coffee pot." Svenson eased into a big leather chair near the fireplace.

"Certainly, these old ears still vork gude."

"Grandpa likes hearing pretty girls' voices better than newscasters, anyway," said Carl Junior, grinning at his father as he twirled the dials.

> "Milk is nature's greatest gift,
> Drink it for your daily lift.
> Topmost milk is pure as gold,
> Vitamin D for young and old."

19

sang a seductive male voice, as some apparently contented cows mooed happily in the background.

"And now, ladies and gentlemen! Here in Bismarck we had six inches more snow today, but the wind is lessening tonight. The Legislature had a busy session. The most important measure which came up for final vote was Referendum Bill 313, introduced by the Chairman of the Tax Committee, Senator Frank Grandin, providing for a new method of taxing farmland. As most of you listening know, our constitution now provides that all farmlands are taxed at the same rate – the big farmer pays the same rate as the small one. Bill 313 would tax the big farmer at a higher rate than the little farmer."

"This change in rate would require an amendment to the constitution. Bill 313 would add these words: 'except that farm and ranchlands when under one ownership may be subject to a progressive graduated tax, with the tax rate increasing with increasing value. The people or the Legislature may divide all property, either real or personal, or both, into classes for taxation purposes, and determine what class or classes of property shall be subject to taxation and what property, if any, shall be exempt.'

"Briefly, folks, all that these legal terms mean is that under the new bill the Legislature could provide that if a man owned 160 acres, for example, he would pay the regular taxes, but if he owned five quarters, the tax on each additional quarter could be increased as much as the Legislature decided. This means that big farmers would pay at a higher rate than little farmers.

"The legislators are all happy about this new proposal. They feel it will put more people on farms, benefit the small businessman through more trade, let those who have the most carry the biggest tax burden and prevent farms from becoming excessively large. This measure now goes on the referendum ballot to be voted on by you people at the June primaries.

19

"Despite the heavy snowfall, hundreds of women were present at the fashion show put on in Bismarck by glamorous models from ...."

"Well," said Christian Ericson, chuckling delightedly, "that sounds good to me – make the big farmers pay more than the little farmers like us. Let's hope it goes through. It'll sure fix that landhog, Nils Rocksvold. Everybody in this part of the state hates him. They'll sure be glad to see him pay through the nose for his piggishness."

"That law ought to help reduce our taxes, too," commented Hilda Olsen, looking up from the bootee she was fashioning for her prospective grandchild. "If there are more small farms, there ought to be a bigger revenue and that will help reduce taxes."

"The proposition will probably carry if the businessmen are for it," added Eric Olsen. "I believe in letting the wealthy pay a good share. Here I am with six children and that Nils Rocksvold hasn't any. It's right he should pay more."

"You know," said Carl Svenson, "it's likely to help my church – more people will mean a bigger congregation for me. I'll probably comment favorably on the idea in my sermon some Sunday soon."

"Carl, my son," spoke Sven Svenson slowly, "your old father vorries. This law I think is not gude. Vy, I don't know yet, but I vill think on it. For this land your father vorked with these hands; all rough they are, for many stones they peek. I leave the ol' country with my teekut but not much food for terty days. Vy I leave the homeland? To get land for my sons!"

In his excitement, the old man struggled to his feet and pointed a gnarled finger at the commentator.

"Now, these politicians at Bismarck say I pay more tax rate on my 320 acres than Olie who have only 160. Loud talk! Everybody pay alike.

"Vy we come to America? To get land. Carl, my son, don't yump when you not know where you're yumping. Don't be silly fools. Tonight I see what says in the Gude Book."

19

FARM BUREAU AND INDIVIDUAL FREEDOM

"Father," begged Helen Svenson, "please don't you worry – we won't vote for anything until we have studied and prayed about it. Help him sit down, Carl."

"Yes, Father," added Carl Svenson tenderly as he patted the old man's arm, "your advice is always wise. Think this all out and let us know what you decide. I've never made a mistake when I've listened to you. We'll think it through and talk it all over again."

SCENE II

At a meeting of Bergen Farm Bureau two weeks later

"In conclusion, members of Bergen Farm Bureau," proclaimed Senator Grandin, raising his arms eloquently, "may I summarize the many advantages of Bill No. 313, which I had the honor of introducing into the Senate of our great state, and which you will be privileged to vote on as a referendum measure in June."

"This law is a real step forward in progressive legislation. It promotes the brotherhood of man and the family-sized farm. It assures more farmers and less landhogs, more customers for business, more family security, a vastly increased production from land and lower taxes for all. It is not strange then, that all men are in favor of this measure – a measure designed to guide the destiny of the farmers of this country – the greatest group of fine citizens that America has yet produced.

"I appreciate your support. I am thankful for your cooperation and I shall long remember you fine people who have welcomed me here tonight."

"As your representative I pledge to you that I will continue to devote my untiring effort, all my strength – God willing – and my very life blood to the interest of you, my constituents. I thank you."

"Our sincere thanks to you, Senator Grandin," said the Farm Bureau president, as the ringing applause died down. "How much time do you wish to allow for answering questions?"

"Oh, – er – well, are questions necessary? I thought I covered everything in my speech."

"Farm Bureau usually allows time at the end of a speech for questions," explained the president.

"In that case," replied the Senator, "I will defer to Farm Bureau's wishes. Who would like to be the first to testify for this measure?"

"Mr. Chairman and Senator," said a member in the third row, "I express what I am sure is the thought of every Farm Bureau member in this hall tonight. This law is badly needed and will do a great deal for the farmers of this state. I suggest that we pledge our heartiest cooperation to Senator Grandin and that we campaign vigorously."

"Mr. Chairman," said Carl Svenson quietly, "I should like to ask the Senator a question."

"Certainly," replied the Senator, beaming affably. "Every question asked and answered will bring more enthusiasm for this law. What is your question?"

"Don't you think it is right for a farmer to have enough land so that he can leave some to each of his children?"

"Children," said the Senator rhetorically, "should look after themselves. How well I remember as a tiny lad rising at 4 A.M. to deliver papers, even in the blizzards of winter. If children are put on their own, there will be no juvenile delinquency such as is rampant at present in our fair land."

"Now don't misunderstand me," he continued, as a little murmur arose. "I love children. I'm just against indulging them by starting them out with money and land. Does that answer your question, my good man?"

"I have two more questions," replied Carl Svenson firmly. "I had always thought that the only purpose of any tax was to pay necessary government expenses. Is that the real purpose of this tax?"

19

"My dear sir," replied the Senator, "this tax proposal does not think only of the problem of raising money. It has far nobler purposes such as reform of land ownership. Taking from those who have and giving it to those who have not; forcing the great landowners to share their property; building more equality and brotherhood."

"Thank you, sir," replied Carl. "In other words, you mean that the purpose of this tax is to reform our present system of farmland ownership?"

"Now – this question is for my father, Sven Svenson, who homesteaded the farm which our family owns. Your proposed bill is a graduated land tax. Has any tax authority ever advocated this principle of a graduated land tax?"

"Yes, indeed," replied the Senator. "The greatest tax authority in the world advocated this plan years ago. You may not have heard of him. He is no longer living."

"What was his name?" called out Lars Ericson.

"Karl Marx."

"Thank you," replied Carl Svenson. "Wasn't Karl Marx a communist?"

"Well – er – no, indeed. He was a socialist. Communists and socialists are quite different. He was the great proponent of equality and brotherhood of man."

"As I understand your proposal and your answers to my questions," Carl continued, raising his voice above the confused murmur of comment throughout the room, "this tax has a hidden meaning. It gives to the Legislature a new control over farmers. This measure destroys the individual liberties of private ownership of property, and the right to save and pass on to our children. Now, all farmers' land is taxed at the same rate. You would change this and give the Legislature the right to decide who must pay more and who must pay less. The thrifty man who accumulates land will be penalized, either by having to pay higher taxes or having his land confiscated if he is unable to pay them.

"As a minister, I claim this is un-Christian, for it has no element of human compassion. It would not consider the man with a big family; the man who fell sick and couldn't farm for a year or two; the farmer who had bad luck and got drowned, or flooded, or hailed out. I didn't recognize this law for what it is – a reform measure to divide up the ownership of farmland, to destroy the right of a farmer to work, to save money and to pass on his land to his children. My father left the old country because the laws there made it impossible for a farmer to live and accumulate land. He came to this country and worked all his life for his land. He saw the danger in this bill. He warned us as I am warning you."

"This legislation is dangerous," continued Carl Svenson, raising his voice over a tide of protests as the members began to grasp his reasoning.

"Senator Grandin," called out a member on the aisle, "our state constitution provides now that all farmland shall be taxed at the same rate. You say that when we give the power to fix different rates for different groups of farmers to the State Legislature, they will use it to benefit the small farmer. How do we know that two years from now they won't fix a rate to benefit the large farmer, or the cattle farmer, or some other group at the expense of all the other farmers? 'The power to tax is the power to destroy.' I ask you, Senator, why should this power be controlled by the whims of the Legislature?"

Wisconsin Farm Bureau providing voter registration information to visitors at Wisconsin Farm Progress Days.

"Mr. Thoren, I am hurt," replied the Senator. "Surely you trust your legislators to use the power to fix differing tax rates for different groups, as a sacred trust."

"Mr. Chairman," shouted Abel Brown from the back row, "now the real facts are beginning to come out. I'm shocked by what I hear. I think it is Sven Svenson and Carl who deserve our thanks instead of the senator."

An eager member jumped up in the second row. "We've been told over and over that we should keep our children on the farm. I have five boys, and I want them to be farmers. I'd like to have enough land to give them a start. It certainly looks to me like this law is against me and my children. It would ...."

"But, sir," interrupted Senator Grandin belligerently, "don't you think the general welfare and the brotherhood of man are more important than one individual's children? Now I regret I must be leaving to address another meeting over at Black Mountain. Thank you for your support, and good night to all."

"Mr. Chairman," called out Olaf Johansen as the Senator strode red-faced toward the exit, "why should the farmers be discriminated against? Like everyone else, we pay income tax. We farmers have already paid a graduated income tax on whatever profit we get off our land. Our land is our capital. It doesn't seem fair to tax both the farmer's income and his capital on a graduated scale. Business people can enlarge all they want and still pay the same rate of taxation, yet the senator, who talks so slick, tells me that if I save and accumulate more land, all I can expect is a higher tax rate. I always thought that in the United States we had the right to earn and save as much property as we could. Aren't thrift and savings still honorable?"

"Carl Svenson, do you want to comment on that?"

"Mr. Chairman, I'll try," replied Carl. "Frankly, this measure is a well-disguised attempt to use the taxing power of the government to divide the wealth by dividing the land. We've never had anything like it in America and, if I can help it, we never will. This tax would discourage all farmers from saving and accumulating more land and would deny them the right to give their children a start in life."

"Under this graduated tax, a man who owns a considerable amount of land would cease to make a profit on his land and might be forced to sell it. This is confiscation. When land is confiscated, it comes off the tax roll and seriously reduces the funds available for schools, townships, counties and the state."

"This so-called progressive law would discourage young people from farming. It would cause higher rentals because the added tax would be passed on to the sharecropper, or tenant farmer.

"Fellow members, many of your relatives, like my father, left the old country because of the unfair tax burden. They came to America because the Constitution of this country guarantees the freedom of the individual. One of these freedoms is the right to own land, to save, to invest and to bequeath to our children. This is why my father saw the evil intent in this legislation when I foolishly thought it good.

"Listen to these words of Lincoln:
> You cannot bring about prosperity by discouraging thrift;
> You cannot strengthen the weak by weakening the strong;
> You cannot help the little man by tearing down the big men;
> You cannot help the poor by destroying the rich."

These was a long silence. Then a dozen members leaped to their feet.

"Mr. Chairman!"

"Mr. Chairman!"

"The chair recognizes Andrew Peterson."

"Mr. Chairman. We certainly had the wool pulled over our eyes. Thanks to the clear thinking of old Sven Svenson, we begin to see the facts.

"I move that Bergen Farm Bureau undertake an intensive campaign in opposition to this measure. And that we send a

19

letter to our state Farm Bureau urging Farm Bureau members in this state – every man, woman and child – to expose the real evil in this bill and fight it. You know what Jefferson said to do when the voters don't understand an issue? He said 'Place before mankind the common sense of the subject.'

"I pledge all of my time until June – and I'm willing to help finance any program necessary – to present the dangers of this proposal to the voters of this state. My wife and son both want to help too."

"I hear about fifty seconds to that motion," said the chairman. "But let's talk one at a time. Bill, will you write down the names of those who are offering money? Hal, you put down the offers of time and ideas. I want to say just this: Let's try to get other organizations to cooperate, but if we must, we Farm Bureau members can, and will, defeat this measure alone. Now Nils, you're first. Tell us how you want to help."

SCENE III

One June night, a few months later, the Ericsons and the Olsens are again in the living room of the Svenson farmhouse, listening to the same news commentator.

"And now," boomed the tense tones of the newscaster as Carl Jr. adjusted the dial, "I bring you fairly complete returns from 1604 precincts. For Governor, incumbent Harold Jensen is leading by a majority of 4,162 votes over his opponent, Steven Waring. For Lieutenant-Governor ...."

"Why doesn't he hurry up?" broke in Mary. "We don't care about all that."

"Have patience, daughter," counseled Carl Svenson. "Each one of us in this room is just as anxious as you are. We've worked and thought of little else for five months. The results tonight mean a lot, not to just us three families, but to everybody in this whole country – even though they may not know it."

FARM BUREAU AND INDIVIDUAL FREEDOM

"And now we come to the hottest issue of them all," proclaimed the newscaster. "The results...."

"Does he mean our Proposition 5?" queried Carl Jr. excitedly.

"Ssh!"

"It's been a long time," the announcer continued, "since any measure has aroused such feeling, particularly among farmers, as has Proposition 5, the referendum measure of the graduated land tax on farms. Here's the verdict – fairly complete returns show an overwhelming defeat for the measure. There is no question but that the major credit for the defeat belongs to members of Farm Bureau.

"And now, instead of the commercial, our sponsors, Topmost Milk, have generously turned over their time to John E. Wilcox, state president of Farm Bureau. Mr. Wilcox will continue talking until additional returns enable us to give you further totals. Mr. Wilcox."

"Thank you, Mr. Roberts. As president of the state Farm Bureau, I rejoice for every citizen of this state at the defeat of Proposition 5. In plain words, the graduated land tax is a socialist-inspired measure to 'reform,' through use of the taxing power, the distribution of wealth and property in this state."

"It is un-American. Had it passed it would have destroyed our concepts of individual liberty and the right to private property."

"The credit for defeating this measure by their votes goes to the voters of this state. But every member of Farm Bureau knows, and I want every listener to know, that what made this victory possible was the philosophy of the men and women who settled our state. This philosophy of individual freedom was made real to every voter by the untiring work of Farm Bureau members. We have faith that if the voters of this state really know the facts about a proposition, they can be trusted to decide that issue wisely. Farm Bureau's faith in the voters of this state was justified.

"One of the outstanding leaders in this fight was Carl Svenson of Bergen Farm Bureau. His father, Sven Svenson, was among

19

the old homesteaders who recognized Proposition 5 for what it is – an attempt to destroy the individual liberty of land owners by redistributing private property. These homesteaders explained the hidden meaning behind this measure and the sons took up the battle."

"Gee, Granddad – Dad – aren't you proud?" interrupted Carl Jr.

"The members of Farm Bureau gave $25,206 to help bring the facts to every voter. They gave time and unstinted effort. The women wrote letters to friends, called thousands of voters on the telephone and registered voters.

"Members spent their evenings calling on neighbors and on every voter who wanted to listen. Spurred by the need to defend the dignity and freedom of the individual, farm leaders put modesty aside and requested permission to speak at meetings of other organizations in every community.

"All this effort brought early results. Fifty percent more voters in this state registered for this primary than ever before.

"Ironically, perhaps the highest compliment to Sven Svenson, to his son Carl, and to the farmers and voters of this state was an involuntary tribute paid by Senator Grandin, the author of Proposition 5, in a press interview just a few minutes ago.

"Senator Grandin stated angrily, 'If it hadn't been for that old Swede and his preacher son and those reactionary farmers who raised such a fuss, this state would have seen a great reform. But we'll wait and try again.'"

Farm Bureau Leadership

Chapter 20

AFBF President Dean Kleckner (far left) and past presidents Charlie Schuman, Allen Grant and Bob Delano (left to right).

(Photo by American Farm Bureau Federation.)

20

DEMAND EXCEEDS SUPPLY

There is an old saying that leaders are born, not made. Like many old sayings, this is true only in part. Seemingly some people do have a marked natural aptitude for leading others. But leadership is fundamentally a skill which is developed by training and by practice. It gains strength through exercise and not overnight.

One horse may be born with more ability to run fast than another horse. Yet without training he will not become a successful race horse. Another may be born with less ability for speed, yet when trained he may become a winner.

Though some persons have more inborn qualities of leadership than others, it is possible, through education and experience, to multiply many times the ability of any individual to lead.

Farm Bureau has thousands of volunteer leaders, and thousands more who are developing into future leaders. But it also has a steadily increasing demand for leaders who can make work in Farm Bureau a happy and satisfying experience for every member.

Farm Bureau is in the leadership business because it is running interference for ideas – community ideas, agricultural ideas, American ideas, world ideas. It needs many types of leaders in its business, with many different abilities: speaker and research people, writers and organizers, young people and old people, children and middle-agers, generators of enthusiasm and wise counselors, dynamic promoters and thoughtful decision makers, planners with vision and doers who know how to get things done.

Farm Bureau's market for leaders never suffers from a surplus. With its constantly widening program and member services, Farm Bureau has an ever-increasing need for trained leaders for today and for leaders in training for tomorrow.

Leadership in Depth

Farm Bureau members elect their official leaders. No higher power appoints them. They are not drafted, they are selected volunteers – and they number in the thousands.

Farm Bureau leaders develop both the ability to cooperate and the ability to lead. Actually, Farm Bureau does not have followers – it has cooperators. There is a vast difference in meaning between those two words. Most sheep are followers – but they are not cooperators.

In Farm Bureau each member needs both abilities, for today's member may be tomorrow's leader and vice versa. No one is either a leader or a cooperator all the time, not even the President of the United States. When he addresses Congress he is a leader – when he talks to his wife he probably is a cooperator. Individuals shift constantly from one role to the other.

Leadership at the top is not enough in Farm Bureau. The organization's power is in its members and in its community groups. Good leadership is vital in every community, every county and every state. It is important that it be spread out and multiplied until every last member of Farm Bureau shares in its benefits.

One reason why the American Army has never been defeated is that the sergeant, in his own group, is as truly a leader as is the general. The success of any army depends in part on the soldiers' ability to obey instantly, yet that ability need not destroy the capacity for leadership. Dwight D. Eisenhower says, "The preservation of both the individual's ability to lead and his ability to work in a team is one of the responsibilities of a good leader."

Two or three strong leaders in a community are not sufficient. If one is unable to perform his duties, there must be others competent to step into his place. Farm Bureau must have leadership in depth.

Leaders with Title

In Farm Bureau the elected officers are the leaders with official titles. But in every group of Farm Bureau, from the smallest community to the largest state, it is easy to pick out other members, without official position or title, who are important leaders nevertheless.

There's the leader who sparks the volunteers. When the presiding officer asks, "Who'll do this job?" this member is always the first to call out, "I'll try" or I'll be glad to do that." You've seen him; he's always there to reach out the first supporting hand when the chairman needs help. Kindled by his example, other volunteers quickly follow.

The trouble-shooter is another leader. Paul smooths out difficulties almost as fast as they arise. If a speaker is misunderstood, it is Paul who jumps up to interpret correctly. "What I think Frank meant ...." If the discussion becomes too heated,

Kurt Inman served on Michigan Farm Bureau's Policy Development Committee representing the State Young Farmer Committee.

(Photo courtesy Michigan Farm Bureau.)

he puts things back in perspective. When the ice is melting rapidly at the July picnic, it is Paul who drives to town for more.

Don't overlook the generator of good will. If someone remarks enthusiastically, "I never attended a more interesting meeting," Helen Hill passes this remark along, not only to the program chair but to the speaker, the chair of refreshments and the presiding officer. After a speech contest, Helen collects the favorable comments on each individual speaker and carefully repeats them to the losing contestants. She makes new members feel welcome and old members feel wanted. She is on no committee, but is a leader in her own right.

There's the leader in judgment. Often before a meeting, you see a group of members gathered around one farmer. "What do you think of it, John?" somebody will ask. "Yes, tell us your opinion," another will urge; and many a matter will be decided when the word is passed around that "John thinks this is a good idea." John's judgment is respected and the members seek his quiet opinion. He says scarcely a word in meetings, but he leads in decision-making because his considered opinions are wise.

Then there's the leader with specialized knowledge. When a question involving money arises, some member is sure to say, "Let's ask Thompson," or "Thompson's our finance man; what's he say?" When there's a problem of raising salaries or planning a new addition to the meeting hall, Thompson is the financial leader to whom members look for guidance.

Invaluable is the creator of new ideas. Whether an original plan is needed for raising money for office furnishings, or a unique way of attracting members, or an unusual feature for a program, Nathan and Bess Blake, working as a team, are always ready with a good suggestion. Members often say, "Let's ask the Blakes. They'll have a good idea."

At one county annual meeting a middle-aged member moved unobtrusively about before the session started, doing all sorts of little odd jobs. He adjusted the windows for better ventilation, introduced some early comers, replaced a light that had burned out. "Are you the handyman, Bill?" someone asked jokingly. "No," he answered twinkling, "I guess I'm the unofficial chairman of the unfilled needs committee."

In every community Farm Bureau and in every county and state, there are these leaders and many more like them: leaders without title, who have not been elected or appointed, but who voluntarily contribute effective leadership. They are working not for recognition or glory, but for the good of Farm Bureau, and each contributes what his special tastes and talents qualify him to do. Actually, Farm Bureau has almost as many potential leaders as it has members.

OPPORTUNITIES FOR LEADERSHIP TRAINING

Farm Bureau annually plants the seeds for a bountiful crop of leaders. Opportunities for training are everywhere. Participation in meetings and serving on committees are some of the best ways for self-training in leadership.

The American Farm Bureau Federation holds annual leadership conferences and institutes. Similar opportunities are offered by most regions, states and counties, and by the Young Farmers and Ranchers Committee. These workshops give practical courses in leadership skills, including a better understanding of Farm Bureau, meeting procedures, policy procedures, committee work, planning, speaking, organizing and other leadership abilities valuable to any member.

QUALITIES OF LEADERSHIP

Effective leaders differ so widely in their personalities, methods and ways of working that it is hard to pick out characteristics common to all of them. Thelma Green is aggressive and inspiring – Henry Black is so mild and self-effacing that people rush to help him. John Jones seems to lead largely by personal example – Bill Smith by persuasion. Mary Jones is businesslike – Sally Smith is informal. One pushes – another invites, yet all are successful.

There seems to be three qualities which most good leaders have in common. These are:

VISION TO PLAN
POWER TO UNITE
COURAGE TO WIN

Vision to plan includes the imagination to sense what others want, even though they themselves may not see it clearly. It means combining the far view and the broad view. It means teaming imagination in seeing an objective with practicality in planning how to reach it.

Vision to plan in Farm Bureau means looking beyond this year's horizon and seeing what will benefit agriculture and the nation in the long run. It means stating your conclusions as a broad, long-term policy recommendation and making practical plans for the development of the policy.

Power to unite means the ability to guide people with diverse ideas to agree upon decisions and to generate the warmth and harmony of true cooperation. It means getting people to rally around an idea and to focus their energy on it.

A leader, if he is to unite people, must inspire confidence in his motives and integrity; draw out the diverse talents of individual members; and combine them in a common effort.

In Farm Bureau the power to unite means persuading your fellow farmers to forget their little differences and work together in support of their great common interests.

Courage to win is born of confidence in yourself and in your aims. It is bravery to act quickly and decisively. It is a lasting determination to win – by detouring around obstacles, plowing through them, or leaping over them. In Farm Bureau, it means fighting staunchly for what you believe is right and fighting on until you are victorious.

LEADERSHIP AND HUMAN RELATIONS

The true leader understands that his success depends not on power over people, but on power with people.

Leadership, to be effective, must get something done. Performance is the ultimate test. But we should judge a leader also on how he gets it done. When the goal is reached, do the members feel imposed upon and disgruntled, or are they happy and eager for another job?

A great leader is a magician working with human elements. He transforms members who just sit, into members who do things. He changes apathy into zeal, doubt into conviction, inertia into initiative and beliefs into action.

The seasoned leader brings human elements together and carefully watches the reactions. She realizes that her power comes from others. She doesn't try to do all of the job herself. Farm Bureau has a well established policy that a good leader chooses her fellow workers carefully, explains the job to be done thoroughly, places the responsibility on their shoulders and then backs them up.

LEADERSHIP MUST BE POSITIVE

Leadership must be dynamic – active, not passive. A political leader is often but a reflector of popular opinion rather than a leader of it. While the positive leader cannot act in opposition to the opinions of the majority, he can inspire them to make a better decision than they would otherwise make.

(Photo by Bob Krauter, California Farm Bureau.)

Lori Sousa, Tulare County Farm Bureau, California, speaks before the state Farm Bureau annual meeting.

It is not enough to have the members agree with you. A great leader must set every member on fire in support of Farm Bureau's program.

Probably at least once in the life of every leader there comes a time when he must risk his leadership in support of his convictions. Listen to one leader who had the courage to place his leadership on the line:

> "There are two opinions about how our Farm Bureau can progress. I have studied both of them, long and prayerfully. I have made up my mind there is only one way I can go – one road down which I can lead.
>
> "You are free to do anything you want: free to change your leader; free to select the road on which you will travel. In fairness, however, I must tell you this - if the majority of you want to go down another road, you should get someone else to lead you. In conscience, I cannot."

The speaker almost invited the members to replace him if they chose a principle which he could not support. That takes courage. What a contrast to the so-called leader who says, "There goes the flock – as their leader I must follow along."

Farm Bureau Today

Farm Bureau has an enviable record of achievement. In relatively few years it has developed into an organization that can count its individual members by millions. It has built a stalwart structure within which the organization works. It has created procedures which protect the freedom and the rights of its individual members in whom the power of the organization eternally rests. It has formulated a basic philosophy by which all proposals are tested. It develops sound, fundamental policies for which it works constantly.

It has cherished and repeatedly defended the integrity of human rights – particularly the right of the individual to freedom and dignity. It has won the admiration and cooperation of legislators and of citizens. It has established its ability to make decisions according to the yardstick of what is best for America.

Farm Bureau has proved its integrity and its courage, its spiritual strength, its deep moral convictions, its will to work and its ability to achieve. It has developed brilliant leaders and attracted an outstanding staff. Farm Bureau is mature, unified and strong.

Farm Bureau remains constantly alert and watchful – for our government is dependent upon the vigilance of citizens in exercising their political rights.

Farm Bureau's Future Lies In Your Hands

Farm Bureau today stands for opportunity, freedom and power.

Opportunity is not something you wait for; it is something you work for. Farmers have built Farm Bureau into the greatest opportunity for every farmer. Maintaining an opportunity is always a challenge.

Farm Bureau is the opportunity and the challenge to be part of the greatest cooperating membership group of agriculture. It is the opportunity to learn, to enjoy and to profit. It is the opportunity to work and to give. It is also a real opportunity to share in shaping a future, rich beyond the imagination of man.

Freedom, too, you have worked for and won. You defend it, or you lose it. Those who founded Farm Bureau built freedom into it – freedom in both its philosophy and its practice. Its members are free to propose ideas, to talk them over, to decide them and to carry them out. They are free to run their organization themselves.

Farm Bureau represents more than freedom within itself; it is the vigilant and aggressive defender of the freedom of all people. Farm Bureau holds personal liberty to be the first essential to human happiness and human dignity. It fights unceasingly to preserve a freedom which gives stimulus to individual energy, intellect and activity. It is convinced that "the greatest glory of a freeborn people is to transmit that freedom to their children."

Power, also, is built slowly and with labor. If it is to endure, power must be honestly acquired and directed to good ends.

Farm Bureau's power is founded on faith in its philosophy of freedom, in the integrity of its aims and in the abilities of its members to work together in unity. Through voluntary cooperation and through voluntary exercise of self-discipline, Farm Bureau has built a power both spiritual and realistic. Farm Bureau's members know that those who hold power should eternally remind themselves that they act in trust.

Farm Bureau members have built an organization in every community that represents opportunity, freedom and power. Even harder work is demanded in the future if these fundamentals are to live and grow.

The basic philosophy which has given Farm Bureau strength will continue to keep it strong. Farm Bureau faces the future with confidence and with courage. This confidence is based on a faith in Christian principles, a faith in its members and a faith in their decisions.

As Farm Bureau grows, its members will grow likewise in wisdom and in ability, in confidence, in loyalty and in devotion.

The best way to prepare for the future is to do the present well.

Through Farm Bureau, you can share in building a better agriculture and in realizing the practical ideal of a world that lives in prosperity, in freedom and in peace. The future lies not in the hands of fate, but in your hands.

AS A MEMBER OF FARM BUREAU, YOU ARE MORE IMPORTANT THAN YOU THINK.

(Montage images courtesy of California, Iowa and Kansas Farm Bureaus.)

The Spirit Grows

BY C. LESLIE CHARLES, 1995

When I am gone
what will people say –
those whose lives I touched
in so many different ways?

When I am gone
the kind of memory
I leave my loved ones
is entirely up to me

When I am gone
my spirit lives, I know;
and all the seeds I planted
will take root and grow.

Appendix A

DEFINITIONS OF PARLIAMENTARY TERMS

Accept an amendment Informal agreement by proposer of a motion to include a proposed amendment; mover may say, "I accept the amendment."

Adjourn To terminate a meeting officially.

Adjourn sine die (without day) An adjournment which terminates a convention or conference.

Adjourned meeting A meeting which is a continuation of a regular or special meeting and which is legally a part of the same meeting.

Adopt To approve, to give effect to.

Adopt a report The formal acceptance of a report. Adoption commits the organization to everything included in the report.

Affirmative vote A "yes" vote to a question before an assembly; an agreement to its acceptance.

Agenda The official list of business to be considered at a meeting or convention.

Amend To change, by adding, deleting, or substituting words or provisions.

Annul To void or cancel an action previously taken.

Appeal A decision of the presiding officer may be appealed from. An appeal requires that the decision be referred to the assembly for its determination by a vote.

Ballot A paper, or a mechanical device, by which votes are recorded. Used to ensure secrecy in voting.

Bylaw A rule of an organization, ranking immediately below the constitution in authority and above the standing rules. Bylaws often include the usual provisions of a constitution.

Candidate One who is nominated or offers himself as a contestant for an office.

Carried Approved by the necessary affirmative vote of the group.

Chair The chairman or presiding officer.

Change in parliamentary situation Phrase used in determining when a motion may be renewed. A change in the parliamentary situation means that motions have been proposed or disposed of, there has been progress in debate, or other changes have occurred to create a new situation so that the assembly might reasonably take a different position on the question.

Changing a vote Request to alter one's own vote which has already been taken.

Charter Written grant of authority, usually from a state to a corporation, guaranteeing rights, franchises, or privileges.

Classification of motions Division of motions into groups, usually according to their purpose or precedence.

Close debate To stop all discussion on a motion and to take a vote on it immediately.

Common law Law developed by court decisions. Judge-made law.

Consideration Deliberation on a subject and examination of it before taking a vote.

Constitution Document containing fundamental law and principles of government adopted by an organized body.

Convene To open a meeting formally.

Convention Assembly of delegates or representatives of allied groups met for a common purpose.

Credentials Certificate or testimonial indicating right of a person to represent a certain group.

Debatable Capable of being discussed.

Debate Discussion or presentation of opinion on a matter pending before a deliberative body.

Delegate Member sent to represent an organized group and empowered to act for it.

Demand The assertion of a parliamentary right.

Dilatory tactics Strategy used to delay action; use of motions and discussion to delay a vote.

Discussion Consideration of a question by oral presentation of views of different persons.

Dispose of motion To remove it from the consideration of the assembly.

Division of assembly A vote taken by counting members, either by rising or by show of hands. Often used to verify a voice vote.

Division of question Separation of a main motion into two or more independent parts, each of which is capable of standing alone.

Ex-officio To hold an office, or position, because of holding another office; e.g., a president may be an ex-officio member of the finance committee.

Executive board Chief committee of an organization. Usually conducts organization business during intervals between meetings.

Expunge To strike out or cancel the record of a previous action.

Filibuster To obstruct or prevent action in an assembly by dilatory tactics, such as speaking merely to consume time.

Floor When recognized formally by the chairman, one is said to have the floor. He is the only person allowed to speak.

Gavel Mallet used by presiding officer of a deliberative body to open and close meetings and to maintain order.

General consent An informal method of disposing of routine and generally favored proposals by the chairman assuming the group's approval, unless objection is raised. Also called "unanimous consent."

Germane Pertaining or relating directly to, having definite bearing upon. Applied to the relationship of amendments to motions.

Hearing Meeting to listen to ideas or arguments with a view to making a decision or recommendation.

Honorary member or officer One who is given membership or office by reason of his eminence or service.

Illegal vote A vote which cannot be counted because it does not comply with the rules of the organization.

In order Correct from a parliamentary standpoint at a given time.

Incidental motions Motions relating to questions which arise incidentally out of the business, or order or manner of considering the business, of an assembly.

Informal consideration A method of considering a question without observing the rules governing formal debate.

Inquiry Question directed to the presiding officer by a member.

Instructions to committee Directions specifying the powers and duties of the committee, the work desired, the type and date of report or similar matters.

Invariable form A motion is said to have an invariable form when it can be stated in only one way and when it is, therefore, not subject to change or amendment.

Irrelevant Not related to, not pertinent, not applicable.

Lay on the table To postpone a motion until a later but as yet undetermined time. Same as postpone temporarily.

Legal vote A valid vote, one which conforms to all legal requirements.

Limit debate To place restrictions on the time to be devoted to debate on a question or the number of speakers or the time allotted each.

Main motion A motion presenting a subject to an assembly for discussion and decision.

Majority vote More than half of the total number legally voting, or if by ballot, more than half of the legal votes cast, unless otherwise defined.

Meeting An assemblage of the members of an organization during which there is no separation of the members except for a recess. A meeting is terminated by an adjournment.

Member in good standing Member who has fulfilled all the obligations required of him by the organization.

Minority Less than half of members or votes. Group having fewer than the number of votes necessary to control the decision.

Minutes Official record of motions presented and actions taken by an organization.

Motion A proposal submitted to an assembly for its consideration and introduced by the words "I move."

Negative vote Adverse vote; vote against a proposition.

New business Any business other than unfinished or old business which may properly be brought before an assembly.

Nomination The formal proposal of a person as a candidate for an office.

Object to consideration To oppose discussion and decision on a main motion.

Oppose To work actively against a measure or candidate.

Order of business The formal program or sequence of different items or classes of business arranged in the order in which they are to be considered by an assembly.

Out of order Not correct from a parliamentary standpoint at the particular time.

Parliamentarian An adviser to the presiding officer and the organization on procedures of organizations; one who is skilled in parliamentary practice.

Parliamentary authority The manual or code adopted by an organization as its official parliamentary guide, which governs in all matters not covered by the constitution, bylaws, and rules of the organization.

Pending question A question, or motion, before the assembly which has not yet been voted upon.

Personal privilege Request by a member for consideration of some matter of concern to himself and related to himself as a member.

Plurality More votes than the number received by any other of three or more opposing candidates or measures. May be less than a majority.

Point of order An assertion, amounting to a demand, addressed to the presiding officer that a mistake should be corrected or a rule enforced.

Postpone definitely To defer consideration of a motion or report until a specific time.

Postpone indefinitely To kill a motion or report by deferring consideration of it indefinitely.

Postpone temporarily To defer consideration of a report or motion until the assembly chooses to take it up again. The old form of the motion was lay on the table.

Precedence The right of prior proposal and consideration of one motion over another, or the order or priority of consideration.

Precedent Something previously done or decided which serves as a guide in similar circumstances. An authoritative example.

Presiding officer Chairman who conducts a meeting.

Previous question Motion to close debate and force immediate vote. Old form of motion to vote immediately.

Privilege of assembly Request for a favor or privilege to be extended to the group.

Privileged motions The class of motions having the highest priority.

Procedural motion A motion which presents a question of procedure as distinguished from a substantive proposition.

Progress in debate Such developments, during the consideration of a question, as might reasonably justify the renewal of a motion.

Proposition A proposal submitting a question of any kind for consideration and action. Includes motions, resolutions, reports, and other kinds of proposals.

Putting the question Submitting a question to vote; taking a vote on a question.

Question Any proposition submitted to an assembly for a decision.

Question of privilege Request or motion affecting the comfort or convenience of the assembly or one of its members.

Quorum Number or proportion of members which must be present at a meeting to enable the assembly to act legally on business.

Recess A short interval or break in a meeting.

Recognition A formal acknowledgment by the chairman indicating that a member has the right to speak.

Reconsider Motion to cancel the effect of a vote so that the question may be reviewed and redecided.

Refer to committee Motion to delegate work to a small group of members for study, decision, or action.

Reference committee Standing committee of a convention to which all motions dealing with a certain subject are referred.

Regular meeting A meeting scheduled in the bylaws and held at definite intervals.

Renew a motion To present the same motion a second or subsequent time at the same meeting.

Repeal To annul or void.

Repeal by implication When two measures are passed which conflict with each other, the portions of the latter motion adopted which conflict with the first, repeal them by implication.

Rescind To repeal, to nullify.

Resolution A formal proposal submitted in writing for action by an assembly. Introduced by the word "Resolved."

Restricted debate Debate which is restricted to the propriety or advisability of a motion in relation to a main motion but which does not open the main motion to debate.

Resume consideration To take up for consideration a motion which has been postponed temporarily. The old form of the motion was take from the table.

Rising vote Vote taken by having members stand.

Roll call Calling names of members in a fixed order as each answers "present" or votes.

Ruling Decision of presiding officer on a question or point of order.

Second An indication of approval of the consideration of a proposed motion.

Special committee A committee appointed to accomplish a particular task and to submit a special report. It ceases to exist when its task is completed.

Special meeting A meeting called to consider certain specific business which must be stated in the call.

Specific main motion A main motion which has a name, a specific form, and is subject to special rules, as opposed to a general main motion. Examples of specific main motions are to rescind, to reconsider, and to resume consideration.

Standing committee A committee to handle all business on a certain subject which may be referred to it, and usually having a term of service corresponding to the term of office of the officers of the organization.

Standing rules Rules formulated and adopted by an organization to meet its own particular needs, and remaining in force until repealed.

Substantive motion A motion which presents a concrete proposal of business; not a procedural motion.

Suppress a motion To kill it without letting it come to a vote.

Suspend Motion to set aside a rule or make it temporarily inoperative.

Teller Member appointed to assist in conducting a vote by ballot.

Tie vote A vote in which the positive and negative are equal, as a 20-to-20 vote. A tie vote is not sufficient to take any action.

Two thirds vote Two-thirds of all legal votes cast.

Unanimous Without any dissenting vote. One adverse vote prevents unanimous approval.

Unanimous consent An informal method of disposing of routine and generally favored motions by the chairman assuming approval of a request for unanimous consent. Is defeated by one objection.

Unfinished business Any business deferred by a motion to postpone to a definite time, or any business which was incomplete when the previous meeting adjourned. Unfinished business has a preferred status at the following meeting.

Viva voice vote A vote taken by calling for "ayes" and "noes" and judged by volume of voice response. Sometimes called "voice vote."

Voluntary organization Nongovernmental organization which members join by choice.

Vote immediately Motion to close debate, shut off subsidiary motions, and take a vote at once.

Well taken A point of order with which the presiding officer agrees is said to be well taken, that is, correct.

Withdraw Motion by a member to remove his motion from consideration by the assembly.

Write in To cast a ballot for a person who has not been nominated by writing in his name.

Yeas and nays Roll-call vote during which each member answers "yea" or "nay" when his name is called.